The Martian Cabal

by

Roman F. Starzl

Double9 BOOKS

The Martian Cabal
by Roman F. Starzl

ISBN: 978-93-59328-76-8

Published by

DOUBLE 9 BOOKS

2/13-B, Ansari Road
Daryaganj, New Delhi – 110002
info@double9books.com
www.double9books.com
Tel. 011-40042856

ABOUT THE AUTHOR

Roman Frederick Starzl (1899-1976) was a writer from the United States. He and his father (John V. Starzl) previously owned the Le Mars Globe-Post newspaper in Le Mars, Iowa. Roman Frederick's father was physician Thomas E. Starzl. His work is largely forgotten today, yet he was dubbed a "master" by E. E. Smith, the pioneer of space opera. Smith's Galactic Patrol may have been influenced by Starzl's Interplanetary Flying Patrol in The Hornets of Space. In Eric Leif Davin's Pioneers of Wonder, Thomas Starzl talks extensively about his father. He was the son of John (born Johann) V. Starzl and Margaret Theisen, and was born Roman Frederick Starzl. The son of Josef Starzl and Magdalena Ruba, John V. Starzl was born on April 9, 1865, in Bischofteinitz (after known as Vvrov) in the Austrian Empire. Josef Starzl moved to the United States in 1878 with his wife and five children (including John). Around the time of his marriage, John V. Starzl sold his pharmacy in Chicago and relocated to Le Mars, Iowa, near where his Bohemian immigrant parents had resided after immigration. He bought the German-language newspaper Der Herold, which became Le Mars Globe Post, and raised Starzl and two other surviving children there.

CONTENTS

CHAPTER I
STRANGE INTRUDER...7

CHAPTER II
SCAR BALTA...12

CHAPTER III
THE PRICE OF MONARCHY...18

CHAPTER IV
TORTURE...23

CHAPTER V
THE WRATH OF TOLTO..28

CHAPTER VI
THE FIGHT IN THE FORT..34

CHAPTER VII
THE FLIGHT OF A PRINCESS...43

CHAPTER VIII
IN THE DESERT..49

CHAPTER IX
PLOT AND COUNTER-PLOT..59

CHAPTER X
ONE THOUSAND TO ONE..66

CHAPTER XI
GIANT AGAINST GIANT..71

CHAPTER XII
"HE MUST BE A MAN OF EARTH"...78

CHAPTER I
STRANGE INTRUDER

Sime Hemingway did not sleep well his first night on Mars. There was no tangible reason why he shouldn't. His bed was soft. He had dined sumptuously, for this hotel's cuisine offered not only Martian delicacies, but drew on Earth and Venus as well.

Sime Hemingway, of the I. F. P., strikes at the insidious interests that are lashing high the war feeling between Earth and Mars.

Yet Sime did not sleep well. He tossed restlessly in the caressing softness of his bed. He turned a knob in the head panel of his bed, tried to yield to the soothing music that seemed to come from nowhere. He turned another knob, watched the marching, playing, whirling of somnolent colors on the domed ceiling of his room.

At last he gave it up. Some sixth sense had him all jumpy. It was not usual for Sime Hemingway to be jumpy. He was one of the coolest heads in the I. F. P., the Interplanetary Flying Police who patrolled the lonely reaches of space and brought man's law to the outermost orbit of the far-flung solar system.

Now he jumped out of bed and examined the fastening of his door, the door to the hotel corridor. There was only one, and it was secure. Windows there were none, and investigation showed that the small ports were all covered with their pivoted safety plates. He extinguished the light, swung aside one of the plates, and peered out into the Martian night. It was moonlight—both Deimos and Phobos were racing across the blue-black sky. The waters of Crystal Canal stretched out before him, seemingly illimitable. Sime knew that the distance to the other side was twenty miles or more. Clear-cut through the thin atmosphere of Mars, he could see the jeweled lights of South Tarog, on the other side.

The hotel grounds, too, were well lighted. Long, luminous tubes, part of the architecture of the buildings, aided the moons, shedding their serene glow on the gentle slope of the red lawns and terraces, the geometrically trimmed shrubs and trees. They were reflected warmly in the dancing waves of the canal, though Sime knew that even in this, the height of the summer season, the outside temperature was very near freezing.

Now a hotel guard came along. He carried at his belt a neuro-pistol, a deadly weapon whose beam would destroy the nervous structure of any living creature. He went past the port with measured stride, and Sime slid back the safety plate with a puzzled frown.

Why was he so nervous? This wasn't the first dangerous mission on which he had embarked in the course of his official duty. And danger was the element that gave zest to his life.

He began a methodical examination of his room, peering under the bed, into closets, a wardrobe. Yet there was no sign of danger. Carefully he inspected his bed for signs of the deadly black mold from Venus that would, once it found lodgment in the pores of a man's skin, inexorably invade his body and in the space of a few hours reduce him to a black, repulsive parody of humanity. But the sheets were unsullied.

Then his gaze fell on the mist-bath. Travelers who have visited Mars are, of course, familiar with this simple device, used to overcome to some extent the exceeding dryness of the red planet's atmosphere. Resembling the steam bath of the ancients, there was just enough room in the cylindrical case for a man to sit inside while his skin was sprayed with vivifying moisture. But his head would project, and there was no head visible.

Nevertheless, so strong was Sime's intuition, he leveled his neuro-pistol at the cabinet and approached. With a sweep of his muscular arm he swung it open—and gasped!

The sight that greeted him was enough to make any man gasp, even one less young and impressionable than Sime. In all of his twenty-five years he had not seen a woman so lovely. Her complexion was the delicate coral pink of the Martian colonials—descendants of the original human settlers who had struggled with, and at last bent to their will, this harsh and inhospitable planet. She was little over five feet tall, although the average Martian is perhaps slightly bigger than his terrestrial cousin. Her hair was dark, like that of most Martians, drawn back from her forehead and fastened at the nape of her neck, from there to fall in an abundant, rippling cascade down her slim, straight back. Her figure was like those delicate and ancient creations of Dresden china to be seen in museums, but elastic, and full of strength. She was dressed in the two-piece garment universally worn by both sexes on Mars—a garment, so historians say, that was called "pyjamas" by our forebears.

And she was defiant. In her hand was a stiletto with long, slim blade. Sime made a darting grasp for her wrist and wrung the weapon from her. It fell to the metal floor with a tinkling clatter.

"And now tell me, young lady, what's the meaning of this?"

Suddenly she smiled.

"I came to warn you, Sime Hemingway." She spoke softly and sweetly, and with effortless dignity.

"You came to warn me?"

"You are in grave danger. Your mission here is known, and powerful enemies are preparing to destroy you."

"You talk like you knew something, kid," Sime admitted. "What is my mission here?"

"You have been sent to Mars by the I. F. P. in the guise of a mining engineer. You are to discover what you can about a suspected plot of interplanetary financiers to plunge the Earth and Mars into a war."

"How so?" Sime asked enigmatically, concealing his dismay at the girl's ready reply. Here was inside information with a vengeance!

"Several shiploads of gray industrial diamonds from Venus have been seized by war vessels carrying the insignia of the Martian atmospheric guard."

Sime nodded. "Go on!"

"Curiously enough, these raids were so timed that they were witnessed by the news telecasters. All of the people on Earth were thus eye-witnesses, and feeling ran high. Am I right?"

"Go on!"

"And of course you know about the raids on the Martian borium mines by pirates armed with modern weapons. In the fights, some of the pirates' weapons were captured. They bore the ordnance marks of the terrestrial government."

"I'm way ahead of you, girlie!" Sime conceded. "Certain financial interests would like to see a war. They're cookin' up these overt acts to get the people all steamed up till they're ready to fight. I'll go further, since you seem to know all about it anyway, and admit that I'm here to find out just who's back of all this. And how does all that tie up with you hiding in my mist-bath with a long and mean lookin' knife?"

The girl dropped her dark lashes in a sidelong glance at the stiletto on the floor. There was a little smile on her lips.

"My usual weapon. Don't you know most of us Martians go armed all the time?"

"Yeh?" Sime grinned skeptically. "And is it a habit of yours to hide in the bedroom of visiting policemen? Come on, kid. I'm going to turn you over to the guard."

For a second it looked as if she would make a dash for the blade glistening there on the floor. But she straightened up, and with a look of infinite scorn said:

"So the mighty policeman of the Sun calls a hotel guard, does he? Please! Believe me, I am myself working for the same object as yourself—the prevention of a horrible war!"

She was pleading now.

"Believe me, you are against forces that you don't understand! I can help you, if you will listen. Let me tell you, the Martian government is itself corrupted. The planetary president, Wilcox, is in alliance with the war party. You will have to fight the police. You will have to fear poison. You will be set upon and killed in the first dark passage. Yet if you help me you may accomplish your object. You must help me!"

"What do you want of me?"

"Help me change our government!"

Sime laughed shortly. He began to suspect that this amazing girl was demented. He thought of the powerfully entrenched rulers of this theoretically republican government. For more than two hundred years, if he remembered rightly, the Martians had been ruled by a small group of rich politicians.

"You propose a revolution?" he asked curiously.

"I propose the return of Princess Sira to the throne!" she declared vehemently. "But enough! Are you going to betray me—I, who have risked much to warn you? Or are you going to let me go?"

Sime looked into her warm, earnest little face. Her lips were parted softly, showing perfect little teeth, and she was breathing quickly, anxiously. Sime was woman hungry, as men of the service often are on the long, lonely trail. He seized her quickly, pressed her little figure to him and kissed her.

For a thrilling instant it seemed that she relaxed. But she tore away, furious, her eyes cold with anger.

"For that," she panted, raging, "you must die!"

She reached the door before he could stop her, and in a trice she was out in the gallery. He raced after her, staring stupidly. Surprisingly, when her escape was assured, she turned back. Her look was still hurt, angry, as she called to him in low tones:

"Look out for Scar Balta, you brute!"

"Who is Scar Balta?" Sime asked himself after locking the door again. The name was not unusual and did not bring any familiar associations to his mind. The given name, Scar, once a nickname, had been in general use for centuries. As for Balta—oh, well—

His mind reverted to the girl again. Her warm, palpitant presence disturbed him.

He composed himself to sleep, strapping his dispatch belt around his waist before crawling into bed. He did not believe that the girl had hidden in his room with murderous intent; rather that she had hoped to inspect and perhaps to steal any papers that he carried. But his last conscious thought of her had nothing to do with her connection with this planet of intrigue, but the soft curve of her throat.

CHAPTER II
SCAR BALTA

Sime breakfasted on one of the juicy Martian tropical pears, and as he dug into the luscious fruit with his spoon he looked about the spacious dining hall, filled with wide-eyed tourists on their first trip to Mars, blissful and oblivious honeymooners, and a sprinkling of local residents and officials.

Through broad windows of thick glass (for on Mars many buildings maintain an atmospheric pressure somewhat higher than the normal outside pressure) could be seen the north banks of the canal, teeming with swift pleasure boats and heavily loaded work barges. Down the long terraces strolled hundreds of people, dressed in garments of vivid colors and sheer materials suitable to the hot and cloudless days. Brilliant insects floated on wide diaphanous wings, waiting to pounce on the opening blossoms.

But the terrestrial agent felt that in this scene of luxury there was a menace. Out of sight, but instantly available, were frightful engines of destruction, waiting to be mobilized against the Earth branch of the human race. And on that distant green planet were people much like these, unconscious still of the butchery into which they were being deftly maneuvered by calculating psychologists, expert war-makers.

His meal completed, Sime sauntered out into the wide, clean streets of North Tarog. He purchased a desert unionall suit, proof against the heat of day and cold of night, and a wide-brimmed Martian pith helmet. Hailing a taxi, he relaxed comfortably in the cushions.

"Nabar mine," he told the driver.

The driver nosed the vehicle up, over the domed roofs of the city and over the harsh desert landscape. The rounded prow cut through the thin air with a faint whistling, and the fair cultivated area along the canal was soon lost to sight.

After half an hour the metal mine sheds grew out of the horizon. But even from a distance of several miles Sime could see that everything was not as it should be. There were no moving white specks of the laborers' white fatigue uniforms against the brown rocks, and no clouds of dust from the borium refuse pile.

The levitator screws of the taxi sank from their high whine to a groan, and the wheels came to the ground before the company office. A man in the Martian army uniform came out. His beetle-browed face was truculent, and his hand rested on the hilt of his neuro-pistol.

"No visitors allowed!" snapped the guard.

"I'm not exactly a visitor," Sime objected, but making no move to get out of the taxi. "I'm an engineer sent here by the board of directors to see why the output of this mine has dropped. Where's Mr. Murray?"

"All settled!" the guard retorted. "Murray's in jail for mismanagement of planetary resources, and the mine's been expropriated to the government. Now, you—off!"

The driver needed no further order from his fare. The taxi leaped into the air and tore back toward the city. It was clear that the military rules of Mars brooked no nonsense from the civilian population, and that the latter were well aware of it.

"Fast work!" Sime said to himself with grudging admiration. Murray was a trusted agent of the terrestrial government. It was he who had first uncovered the war cabal. Sime knew his face well from the stereoscopic service record—a bald, placid man of about forty, a bonafide engineer, a spy with an unbroken record of success, until now. And a fighter who asked no odds, who could manage very well on less than an even break. Well, he was up against something now.

They passed the line of shield-ray projectors, North Tarog's first line of defense against an attack of space, hovered over the teeming streets and parks, and settled on the pavement at the Hotel of the Republic. Sime wanted to go to his room and think things over.

From the concealment of a doorway an officer with a squad of soldiers came up quickly.

"You are under arrest!" said the officer, placing, his hand on Sime's shoulder, while the soldiers rested their hands on their neuro-pistols.

"Would it be asking too much to inquire on what charge?" Sime asked politely.

"Military arrests do not require the filing of charges," the officer explained stiffly. "Come out of there now, Mr. Hemingway."

"I demand to see the terrestrial consul," Sime said, getting out.

"How about my fare?" asked the taxi-driver.

Sime put his hand into his pocket, where he kept a roll of interplanetary script; but the officer restrained him.

"Never mind now," he said ironically. "You are a guest of the government." Then to the driver he added:

"Get on, now! Get on! File your claim at the divisional office."

The driver departed, outwardly meek before the power of the military, and Sime was hustled into an official car. He had little hope that his demand to see the terrestrial consul would be complied with, and this opinion was verified when the car rose into the air and sped over the waters of the canal to South Tarog. It did not pause when it came over the military camps there—the massive ordnance depots in which were stored new and improved killing tools that had long been idle in that irksome interplanetary peace.

They flew on, over the desert, until the Gray Mountains loomed on the horizon. On, over the tumbled rocks, interspersed with the strange red thorny vegetation common in the Martian desert.

Far below them, in a ravine, a cylindrical building was now visible, and toward this the car began to drop. It landed on a level space before the structure. A sliding gate opened, and the car wheeled into a sort of courtyard, protected from the cold of night by an arching roof of glass.

Sime was hustled out and led into an office located on the lower floor of the fortification, or whatever the structure was.

As he saw the man who sat at the desk he gave a startled explanation.

"Colonel Barkins!"

The elderly, white-haired man smiled. He brushed back his hair with a characteristic gesture, and his twinkling blue eyes bored into those of the I. F. P. special officer. The colonel wore the regular uniform of the service; his little skullcap, with the conventionalized sun symbol denoting his rank, was on the table before him. He put out his lean, strong hand.

"Surprised to see me, eh, Hemingway?" he inquired pleasantly.

Sime managed an awkward salute. "I don't quite understand, sir. You gave me my instructions at the Philadelphia space port just before I made the *Pleadisia*. She's the fastest passenger liner in the solar system: I've barely landed here, and it seems you got here before me. It don't seem right!"

Sime watched the colonel narrowly, a vague suspicion in his mind, and he thought he saw a slight flicker in the man's eye when Sime spoke.

But the colonel answered smoothly, with a hint of reproof.

"Never mind questioning me now, Hemingway. The mission is important. I want to know if you remember every detail of what I told you." He nodded to the men, and they filed out of the room. "Repeat your orders."

"Nothing doing, Colonel!" Sime replied promptly and respectfully. "In fact, Colonel, you can go to hell! This is the first time that a man of the I. F. P. has turned traitor, and if your men hadn't so thoughtfully taken my neuro I'd be pleased to finish you right now!"

"But you observe I have a neuro in my hand," remarked the colonel pleasantly, "and so you will remain standing where you are."

So saying, he slipped off the white wig he was wearing, wiped his face so that the brown powder came off, and sat, obviously pleased with the success of his masquerade, useless though it was. He was a typical Martian, dark, sleek-haired, coral-skinned.

"I hate to send a man to his death mystified," said the Martian after a moment, "so I'll explain that I am Scar Balta!"

"Scar Balta!"

"You've heard of me?"

"Uh—yes and no," Sime suddenly remembered the girl of the evening before—the imperious little Martian. She had warned him of Scar Balta.

"If I do say it," said the Martian, "I am the best impersonator in the service of the interests I represent. I did not expect to get information of great value from you, but we do not neglect even the most unpromising leads."

He pressed a button; two Martian soldiers answered promptly.

"Take this man to the cell," Balta ordered. "Provide him with writing materials so that he can write a last message to his family. In the morning take him to the end of the ravine and finish him with your short sword."

"Yes, Colonel!"

"The fellow's a colonel, anyway," Sime thought as they led him away.

They led him downward, along a straight corridor that evidently went far beyond the boundaries of the ravine fortress. In places the walls, adequately lit by the glow-wands the guards carried, were plainly cut out of the solid rock; in others they were masonry, as though the channel were passing through pockets of earth; or—the thought electrified him—through faults or natural caverns.

At last they came to the end. One of the guards unlocked a metal door, motioned his prisoner into the prison cell. A light-wand, badly run down and feeble, with only a few active cells left, gave the only light. As the door slammed behind him, Sime took in the depressing scene.

The stone walls were mildewed, leprous. The only ventilation was through small holes in the door. Chains, fastened to huge staples in the uneven stone floor, with smooth metal wrist and ankle cuffs, were spaced at regular intervals, and musty piles of canal rushes showed where some forgotten prisoner had dragged out his melancholy last days. Sime was glad they had not chained him down. Probably didn't consider it necessary unless there were many prisoners, who might rush the guards.

"Ho, there, sojer!"

The voice was startling, so hearty and natural in this sad place. Sime saw something coming out of a far corner. It was a man in the blouse and trousers of civilian wear; a bald and good-natured man, with a shocking growth of beard.

"Murray's the name," said this apparition with mock ceremony. "And you?"

"I'm Hemingway, Sime Hemingway. Sergeant Sime Hemingway, to be exact. Suppose you'd like to hear my orders?"

"I don't get you," said Murray, shaking hands.

"I mean," Sime explained elaborately, "that I'd like to know if you're Scar Balta, or really Murray, as you say you are."

The other laughed.

"I'm Murray, all right. Feel this scalp. Natural, ain't it? That's one thing Balta won't do—shave off his hair. Too vain. He'd hate to have the Princess Sira see him that way. Ever hear of her? Say, she's a raving beauty. This Balta'd like to be elected planetary president, see—to succeed Wilcox, who has bigger plans. There's always been a strong sentiment for the old monarchy, anyway. The oligarchy never did go big. Follow me?"

"Yeh; go on."

"Well, this Princess Sira has ideas. She wouldn't mind sitting on the throne again. Her great-great-grandpa was jobbed and murdered, and the nobles who did it formed a closed corporation and called it a republican government. So Sira started holding audiences, and gained a lot of power. Among the people—even among some of the nobles.

"Get the idea? Scar Balta is one of the electors. If he married Sira he'd have the backing of the monarchists, and of course he's done a lot for the bosses. They'd elect him to head off the monarchists, anyway. Then heigh-ho for a war with the Earth, to kill off a lot of the kickers—and soft pickins in a lot of ways. Neat, huh?"

"Very neat!" Sime assented drily. "But we won't live to see it. Anyway, I won't. They're going to bump me off in the morning."

"As they have a lot of our men," Murray agreed. "But they won't do it in the morning. Or for several days. Look here!"

He held up his hand. On the back of it was what appeared to be a boil.

"But it isn't a boil," Murray explained. "That was done by a stream of water, fine as a needle, under a thousand pounds pressure. They held it there for a minute at a time—I don't know how many times, because I keeled over. Any time I was willing to give them the information they wanted they'd turn it off. Wasn't important info, either. But what is it to them, how much they make me suffer for a trifle?"

Sime couldn't help the lump that rose in his throat. Men like Murray certainly justified the world's faith in the service.

"Listen, old man," Sime said in a low voice, "out in the corridor—"

But Murray squeezed his hand warningly, pulled him to the floor.

"Might as well get some sleep," the old man said in ordinary tones. "Plenty cool here. Let's lie together."

He kept his hold on Sime's wrist, and, by alternately squeezing and releasing, began to talk in a silent telegraphic code.

"Don't say anything of importance," he spelled out. "They have mikes in here to pick up all we say. Probably infra-red telenses too, so they can see what we do."

So Sime told him, as they huddled together in simulated sleep, about the walled passages, and they speculated on the possibility of felling the guards and breaking their way to freedom through some underground cavern. But at last they slept soundly to await the tortures of the next morning.

CHAPTER III
THE PRICE OF MONARCHY

Had Sime been able to follow and watch the girl he had kissed under such unusual circumstances on the night of his arrival on Mars, he would have been both puzzled and enlightened. After her final warning about Scar Balta she dashed into the luxurious gloom of the passage. At an intersection a maid was awaiting her. She curtseyed as she threw a cape over the girl's shoulder, and together they hurried out into the night.

A magnificently uniformed hotel servant called a private car, drew the vitrine curtains, and saluted as the car lifted sharply into the chilly night air. The car sped across the canal to the jeweled city across the water, to a residence district whose magnificence even the pale night light revealed.

The two women entered a mansion of glittering metal and came to a private apartment.

"Everybody's gone to bed," said the girl, addressing her maid. "That's one thing we can be thankful for."

"Yes, Your Highness. Did you discover anything of importance in the man's room?"

"No. Draw me a bath, Mellie. He—he caught me—and kissed me!"

The maid, with flasks of perfume and aromatic oils in her hand, paused, discreetly impudent.

"You seem not displeased, Your Highness."

"But of that he had no inkling." And Princess Sira laughed. "I left him standing, utterly at a loss. He took me for a common assassin, and yet he wanted to kiss me. That pleased me. But if he had valuable information he kept it. And I promised him death for his kiss."

As Princess Sira, claimant to the throne of a planet, slipped into the tepid waters of her bath, Mellie stood by, her smooth little Martian's face disturbed. For she loved her mistress, and could not comprehend the things she did under ambition's sway.

"Your Highness, couldn't you let your royal friends do these dangerous things for you?"

"For what? For fear? And how could a Martian princess who knows fear lay claim to a throne? No, Mellie, one gets used to it. The enemies of the house of Sira are ever alert. Didn't they murder my father and my mother, and my only brother? My peril in this palace is as great as in the room of a terrestrial detective. Only their fear of the people—"

She was interrupted by the tinkling of a bell. The maid left the alcove, and returned a moment later with the news that Joro, Prince of Hanlon, awaited the princess's pleasure in the ante-room.

"At this hour!" exclaimed the princess. "Did he say what brought him here?"

"Something about a new plot."

"Plots! They fall thicker than rain on Venus. Bid him wait."

Fifteen minutes later, swathed in a trailing orange silk robe that made her look like a Venus orchid, she greeted the prince.

"Greetings, Joro. We seem to have the unusual this night."

The prince, a thin, elderly man of medium stature, smiled admiringly. His sharp features and bright little button eyes gave some hint of the energy which suffused him. Here was a man both ruthless and loyal to his royal house. He addressed her by her given name.

"The hour seems to make no difference with you; Phobos has set, but as long as you are awake there is loveliness enough. I have come, dear one, to tell you that success is ours at last!"

Sira smiled. "I will restrain my joy, my good Joro, until I hear the price."

"Always the same!" Joro chuckled. "A price, 'tis true, but not too heavy, since you are, in a manner, fond of him."

"I've had vague promises from Wilcox," Sira said, with a wry smile. "I would rather trade places with Mellie than be espoused by that madman."

"Not Wilcox, but Scar Balta. He is badly smitten, for which I can not blame him. He has great political power, and the backing of the military. He could have dictated better terms, but for love of you has yielded, point after point. He wants nothing now but your hand in marriage, and is prepared to cede to the royal cause all the advantages he has gained—"

"Not to mention," Sira interjected, "the royal prestige he will gain with the common people."

Joro laughed, a little impatiently.

"True, true! But after all, what does the support of the people amount to? They are powerless. If you are ever to establish your royal house you must have other help."

"And I suppose," Sira continued sweetly, "that you have also arranged a deal with the central banks and the secret war interests?"

Joro coughed uncomfortably.

"As a matter of fact—you see, my dear princess, there are certain commercial interests—transportation, mining, and so forth. They have defied the power of the bankers. They are likely to upset our whole order of society. They need a set-back. And the military men are chafing at their inaction. The war will be ended before too much harm is done, by agreement of the interplanetary bankers. You see—"

"No!" Sira interrupted him coldly. "No! No! No! Oh, I'm sick of the whole thing! I'm sick of the men I know! I hate Scar Balta, and you too. I would rather be the wife of a common interplanetary patrolman than queen of Mars! I withdraw, now!"

Joro, struck by her vehemence, paled. The muscles of his jaw lumped. From a pocket he took a portable disk-radio, an inch in diameter, and spoke a few words. From outside there was a sudden uproar, shouts and curses. The draperies moved, as with an outrush of air caused by the careless handling of an airlock, and the temperature dropped suddenly.

Sira was irresolute only a split second. With a cat-like leap she seized a short sword from the wall, made a lunge at the prince. But Joro, the veteran of many a battle of wits and arms, parried the stroke with the thick barrel of his neuro-pistol, caught the girl's wrist and disarmed her. The screams of the maid went unheeded.

From the other parts of the palace came sounds of struggle, the clashing of sword on sword.

"Sira! Sira!" Joro panted, struggling to hold the girl. "You must give up your impractical ideas! Take the world as it is. Do as I tell you and you'll not be sorry."

"I relinquish my claims!" the girl cried fiercely. "To-morrow I will publicly announce that decision. All my life has been spent feeding that hopeless ambition. Now I will be free!"

"I am loyal to the monarchy," Joro grunted, pinioning her arms at last. "I will guard your interest against yourself."

He began to shout:

"Hendricks, Mervin, Carpender, Nassus! Here, to the princess's chamber."

Several men, after further delay and fighting, responded. They wore civilian blouses and trousers, but there was that something in their alert carriage that proclaimed them trained fighting men. One of them sat down with a grunt on the threshold, holding his hand to a bleeding wound under his armpit. He appeared to be mortally wounded.

Most of the others carried minor wounds, showing that the palace guards had put up a good battle in the sword-play. Both sides had refrained from using the neuro-pistols for fear that the beams, which readily penetrated walls at short range, might injure the princess.

"Let go!" Sira wrenched herself free. "Where is Tolto? Has Tolto turned traitor? How did you get past Tolto?"

"Do not use that ugly word against me. I implore you!" Joro protested. "What we are doing is out of loyalty to the monarchy—not treason. The monarchy is of greater importance than individuals. Consider your duty to the rule of your fathers! As for Tolto—"

He issued a curt command, and there was the sound of movement. Presently four men staggered in, one to each leg, each arm, of the most impressive giant Mars had ever produced—Tolto, to whom there was no god but the one divinity: and Princess Sira was she. Slow of perception, mighty of limb, he had come into her service from some outlying agricultural region of the red planet. His tremendous muscles were hers to command or destroy, as she wished. He would not have consented to this invasion of her home, she knew!

And he had not. Joro had been too wise to try. A dose of *marchlor* in a glass of wine had done what fifty men could not have accomplished by main strength. Tolto was in a drugged sleep.

The Martian Cabal | 21

Joro said: "He isn't hurt. We will simply send him back to his valley, and you, my dear princess, will do your duty to your subjects!"

And there, though he probably did not know it, Prince Joro harked back to the youth of the human race—the compensatory, atavistic principle that gods, rulers, kings, must hold themselves in readiness as sacrifices for the good of their subjects. Joro might have been a tribal high priest invoking their dread rule in the dawn of time. The Martians were, for all their scientific advancement, still the descendants of those prehistoric human savages. Sira knew, instinctively, that the people who loved her would nevertheless approve of Joro's judgment.

CHAPTER IV
TORTURE

When Sime awoke it was to the rattling of the door. Murray stirred. The light was even weaker than before.

"If they offer you a drink, drink hearty!" Murray muttered, sitting up. "I've got an idea it's going to be a hard day."

But they were not offered any water. Instead they were again conducted before Scar Balta, who looked at them morosely. At last he remarked gruffly:

"If you tin sojers weren't so cursed stubborn, you could get yourself a nice berth in the Martian army. Ever consider that?"

"Talk sense!" Sime said contemptously. "If I threw down the service how could you trust me?"

"That'd be easy," Balta rejoined. "Once the I. F. P. finds out you joined us you'd have to stick with us to save your skin."

He laughed at his prisoners' look of surprise.

"Come, come!" he bantered. "You didn't think that I was ignorant of your purpose here? You, Murray; your spying was excellent, I'll admit. You were the first to give away certain plans of ours. Well, well! We don't hold that against you. Wheels within wheels, eh? It would perhaps astonish certain braided gentleman of our high command to learn that I, a mere colonel, control their destinies. As our ancestors would say, it's dog eat dog.

"Now, how about it? I can make a place for you in my organization. It seems to run to secret service, oddly enough. You will be rewarded far beyond anything you could expect in your present career of chasing petty crooks from Mercury to Pluto and back again."

"Is that all?" Murray asked softly, with a bearded grin.

"Oh no. You will turn over to me all the information you can about the I. F. P. helio code. You will name and describe to me each and every plainclothes operative of the service—and you should have an extensive acquaintance."

"Before you answer," Murray said quietly at Sime's side, "let me suggest that you consider what's in store for us—or you—if you don't take up this offer."

"Why, you—" Sime whirled in astonished fury upon his companion. "Didn't you—"

But he did not complete his reference to last night's surreptitious conversation. It seemed that he saw the merest ghost of a flicker in Murray's left eye.

"—Didn't you say you'd stick no matter what they did?" he finished lamely.

Murray hung his head.

"I'm getting along," he muttered. "Not as young as I used to be. This life is getting me nowhere. Why be a fool? Come along with me!"

"Why, you dirty, double-crossing hound!" Sime's exasperation knew no bounds. For an instant he had believed that Murray was enacting a little side-play in the pursuit of a suddenly conceived plan. But he looked so obviously hangdog—so guiltily defiant....

Crack! Sime's fist struck Murray's solid jaw, scraping the skin off his knuckles, but Murray swayed to the blow, sapping its force, and came in to clinch. They rolled on the floor. Murray twisted Sime's head painfully, bit his ear. But in the next split second he was whispering:

"Keep your head, Sime. Can't you see I'm stringing him? Take that!" And he planted a vicious short hook to Sime's midriff.

Balta had squalled orders, and now Martian soldiers were bursting the buttons off their uniforms in the scrimmage to separate the battlers. Bruised and battered, they were dragged apart. Murray's one eye was now authentically closed, and rapidly coloring up. Unsteadily he got to his feet. With mock delicacy he threw a kiss to his late antagonist.

"Farewell, Trueheart!" He bowed ironically, and the men all laughed.

Balta grinned too. "Still the same mind, Hemingway? All right, men, take him up to the observation post. Here, Murray, have a drink."

Sime was led up a seemingly endless circular staircase. After an interminable climb he saw the purplish Martian sky through the glass doors of an airlock. Then they were outside, in the rarefied atmosphere that sorely tried Sime's lungs, still laboring after the fight and long ascent. The Sun, smaller than on Earth but intensely bright, struck down vindictively.

"A good place to see the country," laughed the corporal in charge. "Off with his clothes!"

It was but a matter of seconds to strip Sime's garment from him. They dragged him to an upright post, one of several on the roof, and with his back to the post, tied his wrists behind it with rawhide. His ankles they also tied, and so left him.

It was indeed an excellent point of vantage from which to see the country. The fortress was high enough to clear the nearby cliffs of low elevation, and on all sides the Gray Mountains tumbled to the horizon. To the north, beyond that sharply cut, ragged horizon, lay the big cities, the industrial heart of the planet. To the south, at Sime's back, was the narrow agricultural belt, the region of small seas, of bitter lakes, of controlled irrigation. Here the canals, natural fissures long observed by astronomers and at first believed to be artificial, were actually put to the use specified by ancient conjecture, just as further north they had been preempted as causeways of civilization. Sime painfully worked his way around the post so that he could look south. But here too nothing met his eye but the orange cliffs with their patches of gray lichen. There was no comfort to be had in that desolate landscape. Nevertheless, Sime kept moving around, to keep the post between himself and the Sun. Already it was beginning to scorch his skin uncomfortably.

By the time it was directly overhead Sime had stopped sweating. The dry atmosphere was sucking the moisture out of his body greedily, and his skin was burned red. His suffering was acute.

The Martian day is only a little more than a day on Earth, but to Sime that afternoon seemed like an eternity. Small and vicious, with deadly deliberation, the sun burned its way down a reluctant groove in the purple heavens. Long before it reached the horizon, Sime was almost unconscious. He did not see its sudden dive into the saw-edge of the western mountains— knew only that night had come by the icy whistle of the sunset wind that stirred and moaned for a brief interval among the rocks. The keen, thin wind that first brought relief and then new tortures, to be followed by freezing numbness.

Above, in the blackness, the stars burned malignantly. Drug to his misery they were, those familiar constellations, which are about the only things that look the same on all planets of the solar system. But they were not friendly. They seemed to mock the motionless human figure, so tiny, so inconsequential, that stared at them, numerous tiny pinpricks of light, so remote.

There was no dawn, but after aeons Sime saw the familiar green disk of Earth coming up in the east, one of the brightest stars. Sime fancied he saw the tiny light flick of the moon. There would be a game of blackjack going on somewhere there about now. He groaned. The Sun would not be far behind now.

But he must have slept. The Sun was up before he was aware of it. A man with a caduceus on his blouse collar was holding his wrist, feeling his pulse. He seemed to be a medical officer of the Martian army. His smooth, coral face was serious as he prodded Sime's shriveled tongue.

"Water, quick!" he snapped, —"or he's done for."

His head was tipped back and water poured into his mouth, but Sime could not swallow. The soldier with the bucket poured dutifully, however, almost drowning the helpless man. It helped, anyway; and Sime returned to half-consciousness. A few minutes later, when Scar Balta came to inquire if he had changed his mind, Sime was able to curse thickly. And around noon, when Murray, jauntily dressed in the uniform of a Martian captain, bid him a cheerful good-by, Sime was almost fluent.

His torture had now reached the pitch of exquisite keenness that made it something spiritual. Solicitously they kept him alive, and far back in his mind Sime wondered why they bothered to do that. Couldn't they be satisfied with what they could learn from Murray?

So passed the second day, and the third.

On the fourth day Sime was able to drink water freely, and to eat the food they placed into his mouth, a fact which the medical officer noted. The torture was wearing itself out. Sime's body was emaciated, stringy, burnt black. But his extraordinary toughness was weathering conditions that would kill most men. Balta shook his head in wonderment when this was reported to him.

"Can't wait any longer for him. Must get back to Tarog. You might as well put him out of his misery. By the way, I'm convinced that Murray is double-timing me. But I'll attend to that personally."

From his post of pain Sime saw the official car leave toward Tarog. Had he known of Balta's remark he would not have been puzzled so much by what he saw.

As the ship was about to disappear over the ragged northern horizon, Sime's bleared eyes saw, or he thought they saw, a human figure silhouetted against the pitiless sky. It was a tiny-seeming figure at that distance, but it was clear-cut in the rare atmosphere. Then it plunged from sight.

"Somebody taken for a ride," he muttered, half grateful for the brief distraction from his own misery.

The medical officer, to whom the long climb was arduous, delayed his mission to the roof, and that was why, several hours later, Sime was still alive to see another ship appear to the north. It was large, sumptuous, evidently a private yacht. Its course would bring it within a mile of the fortress, and with sudden wild hope Sime realized that if he were seen he might expect relief. He began to tug at his bonds. They were tough, but they would stretch a little. His haphazard movements had already worn them against the rough post, and now he began to struggle violently. If he could only get his hands loose, he could wave....

The thongs cut into his flesh, but his wrists were numb and swollen, and he did not mind the pain. His muscles stood out hard and sharp, and with a supreme effort, aided by the growing brittleness of the rawhide in the dry atmosphere, he snapped his bonds.

The ship was now quite near, and he waved frantically. He fancied he saw movement back of the pilot ports. Faintly he heard the hum of the levitators. Now it turned—no! It yawed, now toward him, now away, purposelessly, like a ship in distress. It made an abrupt downward plunge that scraped a crag, and just missed a canyon wall.

Again it twisted, came down with a long, twisting motion, struck a rock upside down, slitting a long gash in its skin, clattered to the rocks so close to the fortress that Sime could not see it. Now desperation gave the prisoner superhuman strength. Regardless of the pain, he burst the thongs about his ankles, tottered to the edge of the roof.

There was a battle going on below. Men seemed to be running, shouting. Someone, using a massive plate of metal as a partial shield against the neuro-pistols, was creating havoc. Sime tried to focus his giddy eyes on the scene. It seemed always to be turning to the left, to be circling around him. With tottering steps he tried to follow it, keeping to the brink of that lofty tower— uselessly. Now it was rocking, flying straight toward him, and, gratefully, Sime gave up the struggle, closed his eyes.

CHAPTER V
THE WRATH OF TOLTO

Tolto awoke from his drugged sleep in the cargo room of a pleasure ship. He was thoroughly trussed up, for Prince Joro's servants had a wholesome respect for the giant's strength. Even in his supine position power was evident in every line of his great torso, revealed through great rents in his blouse. His thighs were as big around as an ordinary man's body, and the smooth pink skin of his mighty arms and shoulders rippled with every movement that brought into play the broad, flat bands of muscle underneath.

A chain of beryllium steel was passed around Tolto's waist, and close in front of him the smooth, shining cuffs of steel around his wrist were locked to the chain. Short lengths of chain led to cargo ringbolts in the floor, holding fast Tolto's cuffed ankles.

To anyone looking at Tolto, just then, these extreme precautions might have seemed absurd. Prince Joro, however, was a good judge of men. It would have pleased him best if Tolto had been quietly eased from his sleep into death, but he knew that such a murder would have destroyed forever his chances of winning Sira to his plans. He meant to see Tolto safely and demonstrably returned to his home valley, and in order to accomplish this the more surely, he had him loaded aboard his own ship, and instructed his captain to take the little used desert route.

Tolto lifted his hands as far as he could and looked wonderingly at them. His child-like face, with the soft, agate eyes, expressed only bewilderment. He lifted his voice, a powerful bass.

"Hi, hi! Let Tolto go! The princess may call!"

There was no answer, only the rhythmic hum of the levitators. Again Tolto cried out. But there was no answering sound. The Sun poured in through the ports, and when presently the ship changed its course, the light fell full in his face, almost blinding him. The giant endured this without complaint.

Several hours later, however, his patience snapped, and he roared and bellowed so loudly that a door opened and a frightened face appeared. Back of it was the chromium glitter of the ship's galley.

"Be still, big one!" admonished the cook. "The captain is resting. He will have you chained standing if you disturb him with your bellowing."

"I wanted only to know where I am," Tolto replied, subsiding meekly. "I drank overmuch and some larksters tied me up like this. Release me, so that if the princess calls I may answer."

"The princess will have to call loudly for you to hear," the cook answered jocularly.

"The princess need only whisper for Tolto to hear," the giant boasted, "Come now, shrimp, take these things off!"

"Are you really as dumb as that?" the cook marveled. "Why, sonny boy, the princess couldn't even hear you! Don't you know where you're goin'?"

Vague alarm began to creep over Tolto.

"Where is she?" he asked anxiously. "Isn't she in this ship? Princess Sira never goes anywhere without Tolto. Ask her. Ask anybody."

"The princess may never go anywhere without you, you head of bone," remarked the cook, rather enjoying his own humor, "but *this* time you're going somewhere without her."

"You talk funny talk, but I can't laugh at it. Little bug, tell me now what this is all about, or I will take you between my fingers and squash you!"

The cook's coral face paled almost to white despite himself.

"Listen, big one," he said placatingly. "Have an orange?"

Tolto refused the gift, although he knew this rare and luscious importation from the Earth and was very fond of it.

"Once more I ask you, bug, where is she?"

"Aw, now, listen!" the cook whined. "Don't blame me! I'm only a servant around here. How can I help what they do? Don't glare at me so. Well, she's at Tarog."

"But why—why does she send me away?"

The cook failed to recognize his opportunity to lie in time.

"Well, the fact is—" he hesitated. "The boss—Prince Joro's sending you away. You see, she's going to get hitched up-big important guy. They didn't want you around, bustin' up things every time you turn around. So they're sendin' you back home."

"The princess would not send me home like this," Tolto objected. But he held his peace, and the cook went back to his work, satisfied that he had subdued this dangerous prisoner.

In this he was guilty of no greater error than Prince Joro and the other monarchists. For ages there had been an unfounded opinion that big men are generally slow and stupid. They may often act so, for their great strength serves as a substitute for the quick wit of smaller men. But in Tolto, at all events, this prejudice was wrong. In Tolto's bullet head was a healthy, active brain, and a primitive cunning.

So instead of wasting his strength in vain struggles against the tough steel, he rested, marshalling the facts in his mind.

He utterly rejected the thought that Princess Sira had consented to his removal in this manner, or in any manner. That meant that she was being coerced, and Tolto's eyes grew small and hard at the thought.

Presently he began to test the chains. They were of great hardness and toughness, and so smooth that he could not twist them, for the links slid over one another harmlessly. However, after much quiet effort he found that he could shift his body several inches toward either side of the narrow hold. Here there were a number of locked boxes. One of them, he reasoned, might contain tools.

His closely confined hands were practically useless. He found that he could not reach any of the boxes with his fingers, strain as he might. But he grinned with hope when his head struck one of the handles. His strong teeth closed down on it.

That would have been something to see! The box was of thin, strong metal, but it was heavy. With no other purchase but his teeth, Tolto dragged it to him, on top of him. Now his hands could help a little. He inched it down toward his knees, fearful each moment that a lurch of the ship might precipitate it to the floor with a crash. When his head could push no longer his knees grasped the end of the chest, and managed to pull it down.

Tolto had never heard of the wrestling hold known as the scissors, but he applied it to that box. His mighty sinews cracked under the strain, and stabbing pain tore at his hips. But he persisted, and with a protesting rasp the lid was telescoped inward, breaking the lock.

Breathless, he waited. After minutes he decided that the sound had not attracted attention.

Again he brought his teeth into play, and this time, when the box stood open, Tolto's lips were lacerated by the jagged edges of twisted metal. Triumphantly, he looked inside.

The box contained a set of counterweights for the hydrogen integrator motors.

No bar, nothing that might be utilized to twist off the eyebolts!

Again he set to work. The next box was longer, heavier. It was coated with unpleasantly rancid oil. Tolto's broad chest was covered with blood, partly from gouges in his skin, partly from his crushed lips. But this time he found a bar. It was in the bottom, under some extra valves, but eventually his teeth closed on it, and he fell back, nearly exhausted, for a moment's rest.

He heard a door slam beyond the galley. The words floated out:

"—better go see how he's coming along."

The horrified mate saw the wrecked boxes, the blood-covered giant with a thick steel bar in his teeth, the extra valves scattered about the floor. He whipped out his neuro-pistol, pointed it at Tolto.

But Tolto made no move to resist when the shaken officer gingerly took the bar out of his mouth. He did not move when several shipmen, called by the officer, moved everything out of reach. After half an hour, with many awed comments, they left him alone.

Tolto's battered lips opened in what might have been a grin. Painfully he rolled off the single valve that had been digging into the small of his back. He patiently resumed the tedious task of bringing the valve in reach of his locked hands.

The valve stem was stout, and a foot long. It was just long enough so that Tolto, by lying on his side, could reach one of the eyebolts.

Inserting the stem, Tolto pulled toward him.

The eyebolt turned without resistance. It was free to rotate, and could not be twisted off. A groan escaped from the prisoner.

But in a few moments he tried bending upward. The leverage was highly disadvantageous that way. Still, straining with the last ounce of his strength, he was just able to do it. Pulling down was not so hard.

It took fifty-four motions, up and down, before the tough metal cracked and one chain trailed free.

It was not long afterward that the cook, turning from his work at the electric grill, stared into a face that had once been innocent and peaceful. It seemed the face of a demon.

He would have shrieked, but Tolto took his arm between thumb and forefinger, saying gently:

"Remember, little bug, what I said!"

He was cast, dumb with fear, into the late prisoner's cell.

Tolto had not bothered to remove the chains, but only to twist them apart by means of such tools as he could find to permit free movement of his arms and legs. They dangled from him, tinkling musically.

Now he strode into the main cabin. The ship's crew, having no guests, were playing the part of guests. A man who was shuffling cards, was the first to see him. The cards flew up and showered all over the room.

"He's loose!" this shipman croaked, diving under the table.

"Mr. Yens! Mr. Yens!" shouted the captain, a small, bristling Martian with graying, stiff hair. He snatched the neuro-pistol at his side, pointed it at Tolto, pressed the trigger.

Tolto felt a numbing cold as the ray struck him. But his great body absorbed the weapon's energy to such an extent that he was not killed at once. His flailing arms continued their arc, and one end of chain, whistling through the air, struck the weapon from the officer's hand. Tolto stumbled, recovered. He picked up the pistol and stuck it in his chain belt.

His impulse was to rend, to crush with his hands. The shipmen, except for the officers, were unarmed, and they went down helplessly before the giant fists. Some of them found riot guns, but they might as well have pounded a Plutonian mammoth for all the effect they had on Tolto.

Mr. Yens, the mate, sitting at the controls in the glassed-in cabin forward, turned his head at the captain's cry, and, looking down the short corridor into the main cabin, saw the blood-covered giant coming toward him. Mr. Yens was a brave man; but he had been careless. His neuro-pistol was in his own cabin. He did the best he knew, and snapped the lock.

But Tolto's great bulk smashed in the door as if it were nothing. The unbreakable glass did not splinter, but it bent like sheet metal, and a blow of the giant's fist broke the mate's neck.

The mate had not engaged the gyroscopic control, and immediately the ship began a series of eccentric maneuvers, so sharp and unexpected that no one on board could keep his feet. For a few seconds she straightened, and one of the crew bethought himself of the pistol in the mate's cabin. He sighted on Tolto, clearly visible ahead. Before he could release the ray the ship went into another breath-taking maneuver.

A mountain peak came sliding toward them ominously. They scraped by. The ship dived, throwing Tolto forward, and his instinctive grab threw the elevator up. The levitators screamed madly as they lost their purchase on the air, due to the ship's unstable keel.

"We're goners!" someone shouted. "Kill that fool!"

They bounced off a cliff, turned over and over like a tumbleweed. A cylindrical building, unexpected in this wilderness, loomed up. They seemed about to hit it, but floated past. The rock floor of the valley rushed up. With a crash the ship rolled over, split wide open.

CHAPTER VI
THE FIGHT IN THE FORT

Its coming had been observed. Men wearing the uniforms of the Martian army dashed out, their pistols ready. A man dropped out of a gaping hole in the ship's skin, sat down unsteadily. Others dribbled out.

"Crazy man in there!" one of them shouted. "Look out, he's murderous!" The pistols came up. The soldiers began to close in, showing a certain professional eagerness.

They were perhaps within ten feet when a metal plate, sheared off from the pilot's cabin in the fall, lifted up. Barely visible under it was a pair of large, running feet. One soldier, trying to oppose it with his hands, was knocked senseless and bleeding. He might as well have tried to stop an oncoming rocket ship.

Neuro-pistols, bearing from every side, spanged briskly. They partly neutralized one another. Their charges were partly reflected by the metal and partly absorbed by Tolto's great bulk. He was thoroughly confused now. Every way he looked in this glaring wilderness of desert and rocks were enemies.

But there! An opening loomed, cool and dark. The fortress entrance. Tolto dashed into it. There was the sharp challenge of a guard, unanswered; the futile hiss of a weapon.

The improvised shield wedged on a narrowing stairway. Tolto let it stick, ran up alone. The stairway went round and round, climbing ever higher. The fugitive's lungs were bursting.

At last he came to an airlock. He did not know how to operate it, so smashed through. There was no rush of air, because the pressure had already been equalized in the rush to the wreck at ground level. Panting, listening for pursuers, Tolto looked around.

He found himself on a circular roof, bare except for the airlock and a number of upright posts, whitened by the Sun.

It was some moments before he saw the unconscious figure of a man lying on the very edge of the lofty tower on which he was standing—a man naked and blackened. He was lying on his face, one arm and one foot hanging over space as though he had fallen unconscious at the very edge of the abyss.

Tolto collected his excited wits. This, at least was no enemy. His enemies were in power here. This must be a victim, a possible ally.

The man was stirring. The overhanging arm was feebly trying to grasp something. If he were to roll over—

He did not have time. Tolto dragged him in to the safety of the airlock opening, where he could watch.

There were sounds of pursuit, faint and cautious.

Tolto grinned at the naked stranger.

"Who are you, little bug?" he asked.

Sime Hemingway tried to tell him but his swollen tongue would not behave. Instead, he waved in the general direction of the Sun.

Tolto understood. "From Earth? Good guy, prob'ly. Want this dingus?"

Sime was able to take the neuro-pistol. He knew what was expected of him, and strove to collect his faculties so he could obey orders. He crawled a little way into the lock, where he could be in comparative darkness, setting the little focalizer wheel at the side of the pistol for maximum concentration. Such a beam would require good aiming, being narrow, but if it touched a vital center would be infallibly fatal.

Meanwhile Tolto appraised one of the posts on the roof. It was firmly set in masonry, but he found he could loosen it a little by shaking it. Presently he had it uprooted. It made a splendid battering ram, a war club fit for a giant such as he.

"Here they come!" Sime croaked, and, peering around a corner, took careful aim at the foremost attacker. At the first whispering impact of the beam the Martian sprawled, dead.

The soldiers were caught at a disadvantage. They were expecting club or fist, but not the neuro-beam. Nevertheless Sime had no more easy opportunities. The Martians flung themselves down behind the bulge of the curved stairway, and the air became acrid under the malignant neuro-beams.

None of them reached Sime directly, but the stone walls reflected them to some extent, and even under their greatly weakened power he become cold and sick.

The situation was by no means to his liking. There were other weapons to be reckoned with, and he tried to keep consciousness from slipping away from him. When at last his breathing became easier and his diaphragm moved without pain, Sime knew that danger was greatest. For this relief meant that the Martians had withdrawn down the stairway.

"Good-by, boys!" he thought, as he sprinted up into the comparative safety of the open. He motioned to Tolto, who stood hopefully waiting with his great war club, to stand clear.

There it was! Sime saw the faint phosphorescent reflection against the stone where the stairway curved. He did not wait to see the tiny pellet of the atomic bomb floating up, but threw himself flat on the roof, tugging at Tolto, who understood and followed suit.

Even lying prone, and below the edge of the explosion cone, they were nearly blown off the roof. Though no larger than a pinhead, the bomb had the power of a thousand times its weight in fulminate of mercury. When the rain of small stones and dust had subsided, they rubbed their eyes and saw that the airlock was no more. In its place was a shallow pit, ending with the top of the battered stairway.

"Down after 'em!" Sime husked out of a raw throat. "Before they think it's safe to come after us!"

He led the way, the giant after him, carrying his club and a huge rock fragment. Sime saw a cautious peering head, and that Martian died instantly. Then they were around the bend and in the middle of a fight. Sime deflected a hand that held a pistol, and its beam killed another Martian who was about to let Tolto have it at close range.

There was a light-wand affixed to the wall a trifle further down. Tolto waded through the ruck of smaller men, tore it from its socket and hurled it up the stairs. A short sword bit into Sime's shoulder, but there was no force in the stroke, for in that instant Sime paralyzed his enemy's heart with the beam.

An officer barked a command, and the spang of neuro-beams ceased, to be followed by the lethal rustling of swords. The passage was too crowded for the neuro-pistols, giving the outnumbered prisoners the advantage.

Tolto could not swing his club, but he hurled it, like a battering ram, into the middle of twenty or twenty-five of the garrison who were still below him on the steps, trying to get closer. The heavy timber cleared a lane and the two stumbled down over crushed bodies. Sime was now the only one to use his pistol, for he had no friends there to kill accidentally.

The Martians, were putting up a game battle. They were heirs to the traditions and the spirit of Earth's best fighting men. Science had given them deadly and powerful weapons that could kill over long distances, but they preferred to get close to their adversaries.

But Tolto was a Martian too. He had seized a sword from a dying hand and was wielding it with aptitude and power. No formal thrust and parry for him, but merely a savage sweep that sent swords, arms and heads flying indiscriminately.

Sime, following him, his neuro hissing death from side to side, marveled at his ferocity. He saw a bare-bodied, bleeding fighter leap to Tolto's back, his sword poised for a downward stab for the jugular. Kicking viciously at the man who was just then coming at him, Sime tried to bring Tolto's would-be killer down. But Tolto himself attended to him, dashing him to his death with the elbow of his sword arm.

That diversion nearly cost Sime his life. Fortunately for him he tripped, and the sword-thrust that was to disembowel him merely gashed his side. Sime was beginning to enjoy the fight. The exercise was loosening up his cramped muscles, and the shaky feeling due to the reflected beams of the neuro-pistols was leaving him.

Tolto had smashed down the light-wands as they fought their way down the steps, so that now they were in almost complete darkness. One could still see the occasional rise and fall of a glinting sword and the dark shadow of an arm or head. They were almost clear when Tolto received his first serious wound, a stab in the abdomen that let out a sticky stream of blood.

There was an interval of silence, broken only by the groans of the wounded. The air was thick with the odor of raw blood and pungent with ozone. They had fought their way down perhaps two hundred feet of the stairway, and due to its curve they could see neither top nor bottom.

"I'm stuck!" Tolto muttered.

"Bad?" Sime edged to his side, stepping, in the darkness, on the body of the man who had succeeded in delivering that sword-stroke before Tolto's own blade had cleft him. He felt the edges of the wound, but in the darkness could not tell how serious it was.

"Feel sick? Any retching?" he croaked anxiously.

"Tolto's all right," the giant assured him. "I just said I was stuck."

Sime managed to make a hurried bandage out of the slashed fragment of Tolto's blouse, and again they resumed their descent. Strangely, their enemies further up made no move to attack, although there were many left alive.

Sime laid his hand on Tolto's arm.

"Something wrong here. There's somebody at the bottom of the steps, and the fellows above want to give him elbow room. Well, we'll soon see!"

They crawled up a short distance, began to haul inert bodies down, dragging them as far as the last curve, until they had formed a barricade of nineteen or twenty of their late enemies. It was unpleasant work, but justified by following events.

"Can you just see the loom of it?" Sime asked.

"Yes."

"Watch!"

Sime felt about until he found a small fragment broken from the stone steps. Keeping well within the shelter of the convex wall, he crept toward the bend.

"Dig your fingers into a joint and hold on," he instructed Tolto, locating a crack for himself. Then he tossed the fragment gently over the barricade of bodies.

There was the click of its fall, and a moment later things seemed to turn around. Clinging like leeches to the wall, the two men resisted the warped gravitational drag that would have flung them down upon their waiting enemies below. They seemed to be hanging in a well. Sime had a confused impression of piled-up bodies hurtling down—down.

Thereafter everything was normal again, and they were running down the normal steps. Both had swords in their hands now, and within a hundred feet they were upon the "gravitorser" gun. It was a rather cumbersome weapon, comprising a great deal of electrical apparatus, with a D-solenoid surmounting, whose object was to twist the normal lines of gravitation. It was intended for large-scale operations in the open; the few men remaining below had tried a rather risky experiment, for they might have brought the whole fortress down upon them. Now they were untangling themselves from the corpses that had flown at them as iron flies to a magnet.

Sime and Tolto struck them like a tempest. The light was good and the battle short and sweet. Tolto was slowed up a little, but was irresistible, nevertheless. There is nothing surprising about the seeming immunity of a reckless man in battle. He fights by instinct, taking short-cuts that are not as dangerous as they look because the enemy is not expecting them. So Sime and Tolto fought their way down, until there was no one able to oppose them.

Sime pressed a neuro-pistol into Tolto's hand, warned him to sweep the stairs with it, while he coursed around for some of the pellet bombs. He found them, and two of them closed that avenue of attack with a mass of jumbled ruins.

Now they had a breathing spell. A combination of blind luck and foolhardiness had given them temporary possession of this desert outpost. That was their pawn in the game of life and death—the chance to get back and hide among the millions in the cities of the industrial belt. Certain routine precautions had to be taken. They destroyed the radio apparatus, picked a few days supply of food, threw a couple more bombs and made a search for means of transportation: for there was a desert wilderness of four or five hundred miles to be traversed.

They discovered the egg-shaped hull of an enclosed levitator car in the covered courtyard. It was distinguished by the orange and green stripes which are the Martian army standard. Like all army equipment, it was in excellent condition. The hydrogen gages showed a full supply of fuel.

"We're getting the breaks," Sime crowed to Tolto at they surfeited themselves with water before starting. He had covered his nakedness with an ill-fitting fatigue suit.

"Yeh," Tolto agreed, referring to their numerous wounds with sly humor: "lots of 'em."

Nevertheless, they felt pretty happy when the levitator screws took up their melancholy whine. The rocky valley floor dropped away, and the windowless stone walls of the fortress slid down past them. Now they were even with the top.

Through the ports they could see a group of their late adversaries on the roof, standing in strained attitudes. Their immobility was explained a moment later by an electric blue spark from something in the shadow of their bodies.

Instantly Sime, who was at the controls, threw her hard-a-port, dived, looped up. The first explosion of the tiny projectile tossed them up like a monstrous wave, allowed them to drop sickeningly. The exhaust tubes poured out a dense haze as Sime sought for distance. But they were following him. He was five miles away when they finally got the range. The vessel was jarred as if it had hit a rock. One of the atomic pellets had exploded within a few feet of it. There was a dismaying lurch. Sime picked himself up from the floor and dashed to the controls.

"Everything's all right!" he shouted excitedly.

Tolto, however, was listening anxiously. There was a sharp crackling at the stern, where, in a narrow space, the reaction motors provided the forward motive power. In moments of excitement he referred to himself in the third person. He did so now.

"Tolto's afraid that something's wrong! Smells hot, too!"

"Here, take the wheel!" Sime ordered. The explosions of the shells were becoming less dangerous; they were getting too far away.

Sime burned his hand opening the narrow door. The paint was already blistering off it. The trouble was immediately apparent. One of the integrator chambers, in which atomic hydrogen was integrated to form atomic iron and calcium (sometimes called the Michelson effect), had sprung a leak. The heat escaping into the little room was not the comparatively negligible heat of burning hydrogen, but the cosmic energy of matter in creation. Sime slammed the door. The radiated light was so intense that it stung even his hardened skin.

Looking through the rear range-finding periscope, he saw that they were about twenty miles from the fort. They had ceased firing.

"Won't be long, Tolto," he said, taking over the controls himself again, "before our tail's going to drop off. Got to make time."

It was, in fact, about ten minutes when, without warning, their nose dropped.

"Tail's gone!" Sime announced.

Their momentum, under the destructive rate of speed they had been making, was great, and as the levitators, with independent power supply, still held them up, Sime continued to steer a course for the twin cities of Tarog. He was aided by a light breeze, and the Sun was nearing the western horizon by the time their rate of motion had become negligible.

"Might at well land," Sime decided. "Conserve fuel. If we get a favorable wind to-morrow we can go up and drift with it."

But Tolto, who had been narrowly scanning the terrain, advised continuing a little longer.

"I thought I saw a little smoke, a few miles ahead. Seems to be gone now. But we're still drifting slow."

Sime searched the indicated spot in the ground glass of the forward magnifying periscope. After a few minutes he discovered a blackened spot which might be the remains of a fire. It was surrounded by huge blocks of orange rock, the igneous rock which is the outstanding feature of the Martian desert landscape.

"Looks like he built the fire around there so nobody on the same level would see him," he hazarded. He set the altitude control to fifty feet. There was part of the globular skeleton of a desert hog in the fire; whoever had built it had dined most satisfyingly not long before, and as the fugitives looked their stomachs contracted painfully.

"I could eat a whole one of them myself," Tolto said wistfully.

The urge to descend here was strong upon Sime too. He realized that the fire might have been made by some dangerous criminal—a fugitive from justice; but dangerous men are no novelty to the I. F. P. On the other hand, there was a possibility that it was just some political offender, driven into the desert by persecution. Or a prospector. At any rate, he would have food, or would know where it could be procured.

They had drifted some hundreds of yards farther and the ground was getting constantly more broken, so the best time to land was as soon as possible. Slowly the little ship settled, scraped on a rock and arrested its slight forward motion, crunching solidly in the stony soil.

"Take a neuro, Tolto," Sime advised. "Whoever's here, if he or they are dangerous, we won't get close enough to touch 'em with a sword."

Tolto took the weapon without a word. They locked the door of the ship. Men have been marooned for neglecting that little precaution.

They walked in a spiral course, making an ever-widening circle, looking sharply from left to right. Presently they came to the remains of the fire. The ashes were hotter than the ground, proving that they had been recently made.

But nowhere was there any sign of men. They shouted, but only weird echoes answered.

The ship was now out of sight, and solitude pressed upon them. They felt an uneasy desire to get within comfortable constricting walls.

They found the ship without difficulty.

"Well, whoever it was has lammed," Sime concluded. "Tolto, you climb on top of that rock. Watch me. If you see anybody after me, let 'em have it. I'm going to see if I can scare up a desert hog somewhere."

Neither had stirred from his place, however, before they were suddenly stricken to the ground. They felt the familiar sensation of cold and suffocation—the paralysis caused by a diffused beam from a neuro-pistol. Tolto was a little slower to fall, but he only lasted a second longer. They knew that someone was taking the weapons out of their helpless hands. Then life returned.

"Get up," said a languid voice back of them, "and let's have a look at the looks of ye."

CHAPTER VII
THE FLIGHT OF A PRINCESS

The province of Hanlon, Prince Joro's hereditary domain, began about fifty miles west of South Tarog. It was a region of thorn forests, yielding a wood highly valued for ship-building, and the canal was lined with shipyards, most of which belonged to the prince. The so-called republic had been established before Joro was born, but the reigning family of Hanlon had always been richly endowed with astuteness. Deprived of their feudal holdings by a coup of state, they had won back nearly all they had lost in the fields of finance and trade. Joro was a monarchist for sentimental reasons, not for the profits that might accrue to him.

It was the purity of Joro's devotion to his ideal that made him so dangerous to all who might oppose him. Lesser men might be bribed, frightened, distracted. Not Joro: he believed that the monarchy would soothe the rumblings of internal dissension that continually disturbed the peace and tranquillity of Mars. He drove forward to that consummation with a steadfastness and singleness of purpose such as have carried other fanatics to glory or to the grave. And in addition to his zeal he carried into the struggle his exceptional ability, a knowledge of government and of people.

He had need for all of his rare skill now. It had been an easy matter to carry forcibly the Princess Sira to his palace in Hanlon. Tolto was safely out of the way; Mellie had been dismissed. As for the other palace servants, they had been silenced with bribery or the stiletto.

But Sira had remained adamant, and Joro, abstractedly toying with his laboratory apparatus in the basement of his palace, tried to find the key to her change of heart.

"Can't understand it!" he mused. "She always seemed to have all the royal instincts: cold to suitors, with that delicacy and reserve one finds ideal in a princess. She does all things well, handles a sword nearly as well as I do. Her mind is as keen and limpid as a diamond. She swims like an eel...."

He sighed. "I thought she and I saw eye to eye in this matter. Not more than a week ago she seemed eager for news of the accord I was arranging.

She had no great aversion to Scar Balta. Now she says she will die before she espouses him."

He paused, thought a moment, added, with that absolute fairness and impartiality that was characteristic of him:

"True, Balta is not the ideal prince consort. He would not add kingly qualities to the royal line. But he would confer cunning upon his offspring; and energy—neither to be despised in a royal family that must forever resist intrigue." He sighed again. "The responsibility of king-making is a hard one!"

A sudden thought struck him. "She spoke warmly about the proposed war; could that be at the root of her strange change of heart? After all, she is a woman, and with all her fine, true temper she has a gentle heart. To her the death of a few thousands of her subjects may not outweigh the unhappiness that millions are now experiencing. But the financiers demand the war to consolidate their position, and Wilcox is solidly with them."

With new hope he set down the beaker he was toying with. "Perhaps we can outwit them."

He left the laboratory, climbed a flight of stairs, entered the spacious reception hall. This, like most Martian buildings, was domed. It was richly furnished. The walls were hung with burnished, metallic draperies of gorgeous colors, the floor a lustrous black, the furniture of glittering metal. As the prince entered a servant stepped forward.

"Go at once to the Princess Sira's chamber!" Joro commanded sharply. "Request her to come here. Tell her I have thought of the solution to our difficulty."

Impatiently he paced up and down, stopping at a window for a moment and looking out into the night.

"Your Highness! Your Highness!" The servant was sobbing with excitement. "Your Highness, Princess Sira has escaped!"

Joro left the man babbling, dashed up the broad stairs, unheeding the servants who scattered before him. Their punishment could wait. Just inside the princess's chamber, still unconscious from a blow on the head, lay the guard whose duty it had been to stand before that door. How long ago had she gone? Probably not more than a few minutes.

Joro saw to it that her start would not be much longer. In a few seconds men and women were scouring the palace grounds, and radio orders to the provincial police of Hanlon were crowding the ether.

Sira had contrived her escape without any particular plan in mind. In fact, it had been initiated on impulse. The fellow on guard at her door had excited intense dislike in her. High-strung, and excited by her kidnaping, she had been further annoyed by his officiousness, his fawning, which thinly disguised impudence. The third or fourth time that he intruded on her privacy to ask if she wanted anything she was ready, with the heavy leg, unscrewed from a chair. She felled him in the middle of a smirk, and seized the opportunity created.

It happened that there was a service corridor close at hand. Down this she sped, into the darkness of a boat-house. The doors were barred and locked, of course, but the depths of the water showed a faint greenish glimmer of light. Sira dived in, unhesitatingly, and after an easy underwater swim she emerged in the open canal. There was a considerable swell, for there was a slight breeze blowing from the north across twenty miles of water, but this did not distress Sira at all. She undulated through the waves with perfect comfort. Phobos was just rising in the west, and orientating herself by this tiny moon she struck out in a north-easterly direction, seeking a favorable current to carry her toward Tarog.

Early explorers on Mars were astonished to find that the canals were not stagnant bodies of water, but possessed currents, induced by wind, by evaporation, and the influx of fresh water from the polar ice caps.

This was near the equator, however, and the water was not unreasonably cold, although the night air was, as usual, chilly. After a few minutes Sira discarded her clothing, and so settled down to a long swim.

Ten miles out she struck a brisk easterly current, flowing toward Tarog, and she gave herself up to it. Floating on her back she saw the lights of the prince's ships flying back and forth over the water in search of her—or her body. But none came near her, and she was content.

The abrupt tropical dawn found her in mid-canal, half-way to Tarog. She had no intention of swimming all the way to the capital city, to be fished ignominiously out of the canal by the police. She was in need, not only of clothing, but of clothing that would disguise her. Her coral pink body near the surface of the water would attract attention for considerable distance, and would lead to unwelcome inquiries.

She was glad when she saw a fishing scow anchored in the current ahead of her. The man who owned it had his back to her, fishing down-current. She approached the boat silently and worked her way around it by holding to the gunwale.

Sira now saw that the fisherman was old, gnarled and sunburned so dark that he was almost black, despite the dilapidated and dirty pith helmet he was wearing. His lumpish face was deeply seamed and wrinkled. His sunken mouth told of missing teeth, and his long, unkempt hair was bleached to a dirty gray.

"Have you an old coat you can lend me?" Sira asked, swimming into view.

The rheumy eyes rolled, settled on the water nymph. The old man showed no surprise, but pious disgust. His eyes rolled up, and in a cracked voice intoned:

"Wicked, wicked! O great Pantheus, thy temptations are great—thy visions tormenting. In my old age must I ever and ever live over—"

"Foolish old man!" Sira snapped. "I'm not a vision!" She dragged down an old sack that hung over the gunwale, washed it, and tearing holes in the rotten fabric for her arms and head, slipped it on. It was a large sack, coming to her knees; satisfied, she climbed aboard, where she spread her black hair to dry.

"Not a vision?" the old man quavered. "Then thou art reality, come to gladden my old age—nay—to return youth to me! In my hut there is an old hag. She shall go—"

Sira did not answer. She was neither disgusted nor amused by the dark torrent that stirred in this decrepit old fisherman. She saw only that he had pulled in his nets and was bowing his long arms to the oars, pulling for shore.

It took about two hours before they reached the fisherman's hut, a nondescript, low-ceilinged shelter of logs, driftwood and untarnished metal plates off some wreck. Several times they were hailed by other fishermen, who addressed the old man as "Deacon" and asked jocularly about what kind of a fish he had there.

The deacon's wife awaited them. The old man's description of her as a hag had not been far wrong. She, was as diminutive and weakened as he was ponderous and heavy. She was acid. Her skin was like a pickled apple's; her expression sour, her voice sharp.

"Hoy there, you old hypocrite!" she hailed when they came in earshot. "So this is the way you lose a day! Who's the hussy with you?"

The deacon nosed the old and evil-smelling scow into the bank. His eyes rolled piously.

"The great Pantheus sent her. He said—"

The old woman came closer and inspected Sira, who endured her gaze calmly. That look was like the bite of acid that reveals the structure of crystal in metals.

"Why, she's a lady!" she exclaimed then. "Not fittin' to be on the same canal with you! Come in, my dear. You must be nearly dead!"

She conducted Sira into the hut, which was far neater and cleaner than its exterior suggested.

"A lady!" she repeated. "In that heat! Young woman, what made you do it? Look at those arms—near burnt! Let me take off that old sack. But wait!"

She tip-toed to the door, threw back the faded curtain sharply. The deacon, too surprised to move, was standing there in the attitude of one who seeks to see and hear at the same time. He lingered long enough to receive two resounding slaps before fleeing to his boat, followed by a string of curdling remarks.

Back inside, she proceeded to anoint Sira's body, exclaiming her pleasure at its perfection. The oil smelled fishy, but it was soothing, and it was not long before the claimant to the throne of Mars was deep in restful slumber.

Late that afternoon the deacon returned and hung his nets up to dry. He was dour, his fever having left him. But he had a strange story to impart.

"I think that girl I picked up is the Princess Sira," he told the old woman. "On the fish buyer's barge, in the teletabloid machine, I saw the forecast of her wedding to Scar Balta. And I'll swear it's the same girl!"

"And why," queried his wife, "would she be swimming in the middle of the canal if she was getting ready to marry Scar Balta?"

"That's just it!" the deacon exclaimed, and his eyes began to roll again. "They say it's not a love match! Oh, not in the teletabloid! They wouldn't dare hint such a thing. But the men on the barge. They say there's a rumor that she ran away. And she looks like the girl I picked up, though I thought—"

"Never mind what you thought!" she snapped. "It may be, I served the oligarchy and the noble houses—before I was fool enough to run away with a no-good fisherman—and I can see she is a lady. Well, she can trust in me."

"They say," the deacon hinted, "that if one went to Tarog, and inquired at the proper place, there would be a reward."

The little old woman chilled him, she looked so deadly.

"Deacon Homms!" she hissed. "If you sell this poor little girl to Scar Balta, your hypocritical white eyes will never roll again, because I'll tear them out and feed them to the fish. Understand?"

Considerably shaken, the deacon said he understood.

But the next morning, on the placid bosom of the canal, he forgot her warning. The fleshpots of Tarog called him. Tarog, where he had spent youth and money with a lavish hand. Tarog, where a reward awaited him.

He hauled in his anchor, gave the unwieldy boat to the current and bent to the oars.

Back in the hut, unsuspecting of treachery, Mrs. Homms and Sira were rapidly striking up a friendship. A shrewd judge, of character herself, Sira did not hesitate to admit her identity, and without any prying questioning the old woman soon had the whole story. It thrilled her, this review of the life she had once seen as a servant.

"I wonder if I will ever see Tarog again!" she sighed wistfully.

"You shall!" Sira promised, "if you help me."

"I will do what I can gladly."

"I need a workingman's trousers and blouse, and a sun-hat that will shade my face. I have a plan, but I must get to Tarog. Can you get me these things?"

"I have no money, but wait!" She rummaged with gnarled fingers in a chink in the wall, withdrew a small brooch-pin of gold, with a pink terrestrial pearl in its center.

"My last mistress gave me this," she said smiling sadly. "I will row to the trading boat and buy what you need. There will be a little money left to buy your passage on a freight barge."

And that was why, when the deacon arrived at the head of a squad of soldiers that evening, there was no girl of any description to be found. Ignoring the cowering and unhappy reward seeker, the old woman delivered her dictum to the sergeant in charge.

"Princess? Ha! The deacon, sees princesses and mermaids in every mud bank. His imagination grew too and crowded out his conscience. No, mister, there ain't any princess here."

CHAPTER VIII
IN THE DESERT

Mellie, Sira's personal maid, was too disturbed by her mistress's kidnaping to seek other employment. She saw the teletabloid forecasts of the wedding, made life-like by clever technical faking, but rumors of the princess' escape were circulating freely despite a rigid censorship. She imagined that lovely body down in the muck of the canal, crawled over by slimy things, and she was sick with horror.

Mellie lived with her brother, Wasil Hopspur, and her aged mother. Wasil was an accomplished technician in the service of the Interplanetary Radio and Television Co., and his income was ample to provide a better than average home on the desert margin of South Tarog. Here Mellie sat in the glass-roofed garden, staring moodily at the luxuriant vegetation.

She looked abstractedly at the young man coming down the garden walk, annoyed by the disturbance. There was something familiar in the sway of his hips as he walked.

And then she flew up the path. Her arms went around the visitor, and Mellie, the maid, and Princess Sira kissed.

Mellie was immediately confused. A terrible breach of etiquette, this. But Sira laughed.

"Never mind, Mellie. It is good for me, a fugitive, to find a home. Will you keep me here?"

"Will I?" Mellie poured into these words all her adoration.

"Mellie, the time has come for action. Not for the monarchy. I am sick of my claims. I would give it all—You remember the young officer of the I. F. P.? The one who kissed me?"

"Yes."

"Well, that comes later. First I must consider the war conspiracy. Have you heard of it?"

"There are rumors."

"They are true. Will Wasil help me?"

"He has worshiped you, my princess, ever since the time I let him help me serve you at the games."

"One more question." Sira's eyes were soft and misty. "My dear Mellie, you realize that I may be trailed here? What may happen to you?"

"Yes, my princess. And I don't care!"

As Murray parted from his brother-in-arms, Sime Hemingway, on the roof of the cylindrical fortress in the Gray Mountains, he felt the latter's look of bitter contempt keenly. He longed bitterly to give Sime some hint, some assurance, but dared not, for Scar Balta's cynical smile somehow suggested that he could look through men and read what was in their hearts. So Murray played out his renegade part to the last detail, even forcing his thoughts into the role that he had assumed in order that some unregarded detail should not give him away. He convinced the other I. F. P. man, anyway.

But Murray had an uneasy feeling that Balta was laughing at him, and when the shifty soldier politician invited him into his ship for the ride back to Tarog, Murray had a compelling intuition that he would not be in a position to step out of the ship when it landed on the parkway of Scar Balta's hotel.

Having infinite trust in his intuitions, Murray thereupon made certain plans of his own.

He noted that the ship, which was far more luxurious than one would expect a mere army colonel to own, had a trap-door in the floor of the main salon. Murray pondered over the purpose of this trap. He could not assign any practical use for it, in the ordinary use of the ship.

But he could not escape the conviction that it would be a splendid way to get rid of an undesirable passenger. Dropped through that trap-door a man's body would have an uninterrupted fall until it smashed on the rocks below.

Murray then examined the neuro-pistol that had been given him. It looked all right. But when he broke the seal and unscrewed the little glass tube in the butt, he discovered that it was empty. The gray, synthetic radio-active material from which it drew its power had been removed.

Murray grinned at this discovery, without mirth. It was conclusive.

At the first opportunity he jostled one of the soldiers, knocking his neuro-pistol to the floor—his own, too. And when he apologetically stooped and retrieved them the mollified soldier had the one with the empty magazine.

So far, so good. Murray noted that the wall receptacles were all provided with parachutes. It would be simple to take one of these, make a long count, and be on the ground before he was missed. Provided that he could leave unobserved.

The ship was now well in the air, and beginning to move away from the fort. But they were only ten miles away, and Murray had hardly expected that Balta would be in such a hurry.

"You get off here!" Balta said, and Murray felt the muzzle of the neuro-pistol on his spinal column.

A grinning soldier seized a countersunk ring and raised the trap-door.

"So you're going to murder me," Murray said, speaking calmly.

"I take no chances," was Balta's short answer. "Step!"

Murray stepped, swaying like a man in deadly fear. He lowered his feet through the hole. Looking down, he saw that they were about to pass over a bitter salt lake, occasionally found in the Martian desert. He looked up into the muzzle of the menacing neuro-pistol.

"Balta, you're a dog!" he stated coldly.

"A live dog, anyway," the other remarked with a twisted grin. "You know the saying about dead lions."

Murray's fingers clenched on the edge of the rug. It was thin and strong, woven of fine metal threads. They were just over the edge of the salt lake.

Murray dropped through, but retained his death-like grip on the rug. It followed jerkily, as the men above tripped, fell, and rolled desperately clear.

Murray's heart nearly stopped as he fell the first thousand feet. The rug, sheer as the finest silk, failed to catch the wind. It ran out like a thin rivulet of metal, following Murray in his unchecked drop.

But he had a number of seconds more to fall, and he occupied the time left to him. He fumbled for corners, found two, lost precious time looking for the others. He had three corners wrapped around one hand when the wind finally caught the sheer fabric, bellied it out with a sharp crack. The sudden deceleration nearly jerked his arm out.

Even so, he was still falling at a fearful rate. The free corner was trailing and snapping spitefully, and the greasy white waters of the lake were rushing up!

At any rate, the rug held him upright, so that he did not strike the water flat. His toes clove the water like an arrow, and the rug was torn from his grasp. The water crashed together over his head with stunning force. After that it seemed to Murray that he didn't care. It didn't matter that his eyes stung—that his throat was filled with bitter alkali. All of his sensations merged in an all-pervading, comfortable warmth. There was a feeling of flowing blackness, of time standing still.

Murray's return to consciousness was far less pleasant. His entire body was a crying pain: every internal organ that he knew of harbored an ache of its own. He groaned, and by that token knew that he was breathing.

As unwillingly he struggled back to consciousness he realized that he was inside a rock cave, lying on a thin, folded fabric that might well be the rug that had served as an emergency parachute. He could see the irregular arch of the cave opening, could catch hints of rough stone on the interior.

He sat up with an effort. There was a vile taste in his mouth, and he looked around for something to drink. There was a desert water bottle standing on the floor beside him. That meant he had been found and rescued by some Martian desert rat who had probably witnessed his fall. He rinsed out his mouth with clean, sweet spring water from the bottle, drank freely. His stomach promptly took advantage of the opportunity to clear itself of the alkali, and Murray, controlling his desire to vomit, crawled outside into the blinding light of the Martian afternoon. He saw that the cave was high up on the side of one of the more prominent cliffs. There were many such hollowed places, indicating that the sloping shelf on which he now lay had once been the beach of a vast sea which at some time must have covered all but the higher peaks of the Gray Mountains. It was, of course, the sea that had deposited the scanty soil which here and there covered the rocks. During geologic ages it shrunk until it all but disappeared, leaving only a few small and bitter lakes in unexpected pockets.

There was a succession of prehistoric beaches below Murray's vantage point, marking each temporary sea level, giving the mountain a terraced appearance. A thousand feet below was the white lake, sluggish and dead.

Murray was looking for the man who had saved him. He was able to discern him, after a little effort, toiling up the steep slopes. He was still nearly all the way down. He could see only that he seemed to be dressed in white desert trousers and blouse, and that he wore a broad-brimmed sun helmet. He was carrying something in a bag over his shoulder. He was making the difficult ascent with practiced ease, his body thrown well forward, making fast time for such an apparently deliberate gait.

The desert glare hurt Murray's eyes. He closed them and fell asleep. He awoke to the shaking of his shoulder, looked up into a black-bearded face, a beard as fierce and luxuriant as his own. But where Murray was bald, this man's hair was as thick and black as his beard. He had thrown off his helmet, so that his massive head was outlined against the sky. His torso was thick, his shoulders broad. Large, intelligent eyes and brilliant coral skin proclaimed the man to be a native of Mars.

The man's white teeth flashed brilliantly when he spoke.

"Feeling better? Man, you can feel good to be here at all! Time and again have I seen Scar Balta drop 'em into that lake, but you're the first one ever to break the surface again. He gave you a break, though. First time he ever gave anybody as much as a pocket handkerchief to ease his fall. That lake is useful to Scar. It keeps the bodies he gives it, and none ever turn up for evidence."

Murray was still struggling with nausea. "Want to thank you," he managed. "I got it bad enough. Ow! I feel sick!"

The Martian bestirred himself. He scraped up the ancient shingle, making a little pillow of sand for Murray's head. The Sun was already nearing the western horizon, and its heat was no longer excessive. Murray watched through half-closed lids as the big man descended a short distance, returning with an armful of short, greasy shrubs. He broke the shrub into bits, made a neat stack; stacked a larger ring of fuel around this, until he had a flat conical pile about eight inches high and two feet in diameter.

From a pocket safe he procured a tiny fire pellet. This he moistened with saliva and quickly dropped into the center of his fuel stack. The pellet began to glow fiercely, throwing off an intense heat. In a few seconds the fuel caught, burning briskly and without smoke.

"Wouldn't dare do this in the open," the Martian explained, "if this stuff gave off any smoke at all. The pulpwood mounds down in the flats make a nice fire, but they smoke and leave black ashes, easy to see from the sky. Now you just rest easy. You'll feel better soon as you get some skitties under your belt."

The skitties proved to be a species of quasi-shellfish, possessing hemispherical houses. In lieu of the other half of their shell they attached themselves to sedimentary rocks. They were the only form of life that had been able to adapt themselves to the chemicalization of the ancient sea-remnant. The Martian had left them thin flakes of rock. Now he placed the shells in the red-hot coals, and in a very short time the skitties were turning out, crisp and appetizing. Following his host's example, Murray speared one with the point of his stiletto, blew on it to cool it. It proved to be delicious, although just a trifle salty.

"Drink plenty water with it," the Martian advised him. "Plenty more about five hundred feet down. Artesian spring there. Fact is, that's all that keeps that lake from drying up. You ought to see the mist rise at night."

Murray ate four of the skitties. Then, because the sun was getting ready to plop down, they carefully extinguished the fire, scattering the ashes. The I. F. P. agent felt greatly strengthened by his meal and assisted his host with the evening chores. Nightfall found them in their darkened cave, ready for an evening's yarning.

"I took the liberty of examining your effects,» the Martian began. «Sort of introduced you to myself. The fact that you wore the Martian army uniform was no fine recommendation to me, though I once wore it myself. Your weapons I hid, except for the knife you needed to eat. But you'll find them in that little hollow right over your head. The fact that you're an enemy of Scar Balta is enough for the present. That alone is repayment for the labor of carrying you up all this way."

Murray then told him of work on Mars. There was no use concealing anything from one who was obviously a fellow fugitive, and who might be persuaded to do away with his guest, should he have strong enough suspicions. He told of the war cabal, of the financial-political oligarchy and its opposing monarchists. He related his own discovery and arrest; the pretended enlistment in Scar Balta's forces which terminated in Scar's prompt and ruthless action. When he finished he sensed that he had made a deep impression on his host. The latter spoke.

"What you have told me, Murray, relieves me very much," he said. "I know that we can work together. You might as well know how I came to be here. Perhaps I look forty or fifty years old. Well, I'm thirty. I was news director for the televisor corporations. I didn't have to be very smart to realize that a lot of the stuff we were ordered to send out was propaganda, pure and simple. Propaganda for the war interests, propaganda for the financiers. Commercial propaganda too.

"Why, the stuff we put out was a crime! The service to the teletabloids was the worst. You know how they outstrip the news; hired actors take the part of personages in the news. Ever watch 'em? The way they enact a murder is good, isn't it?"

"We got orders to bear down on your service too, the I. F. P. Your crew has too many points of contact, hiking from planet to planet. The high command couldn't see things the bankers liked, I guess.

"So whenever a man of the I. F. P. figured in the news we always gave him the worst of it. We hired bums to play his part, criminals, vicious degenerates. People believe what they see—that's the idea. I had seen very few of your men but I knew we were giving them a dirty deal. Orders were orders, though. We got lots of orders we didn't understand. Then secret deals were made, and those orders countermanded.

"But the order against the I. F. P. remained standing, and we certainly did effective work against 'em. The people had no way of knowing the difference, either, for the company controls all means of communication, and the I. F. P. does most of its work in out of the way places. Why just to show you how effective our work was—the people, in a special plebiscite, voted to withdraw their support from the Plutonian campaign! But that was going too far; the financiers quietly reversed that.

"At the same time, we got orders to glorify Wilcox, the planetary president. It was Wilcox signing a bill to feed the hungry—after their property had been stripped by the taxes. It was Wilcox the benevolent; Wilcox the superman. Wilcox, in carefully rehearsed dramatic situations, reproduced on the stereo-screens in every home. You know who put over the slogan, 'Wilcox, the Solar Savior?' We did it. It was easy!" He laughed shortly.

"The only time we failed was, when they wanted to end, once and for all, the prestige of the royal house. That was after they had bought the assassination of the claimant, his wife and their son. Didn't dare take Princess Sira too, because she has always been a popular darling. It would have been too raw, wiping out the whole family. They left one claimant, see? And then put it up to us to discredit her!

"Man! That fell down! The first attempt was very smooth, at that. But it brought in such a storm of condemnation they had to drop that.

"You can guess how we boys at the central office felt about it. No wonder we got cynical and lost all self-respect. We couldn't have stood it at all, but sometimes we'd put on a special party, just to let off steam. Did we rip 'em up high and handsome? The more outrageous the flattery we sent out, disguised as news, the more baldly truthful we were in those early morning rehearsals, with the mikes and telegs dead. Wilcox was our special meat.

"Of course, it was foolhardy. One night a mixer in the room below us got his numbers mixed, killing a banquet program on a trunk channel and sending our outrageous burlesque out instead. When the poor fellow discovered his mistake he made for the bottom of the canal. As for me, I made for the desert. I never heard what became of the others, and that was six years ago. I wonder if I've changed much."

The Martian Cabal | 55

"What's your name?" Murray asked suddenly.

"Tuman. Nay Tuman."

"The others must have been caught. As for yourself, orders have been sent all over the solar system to kill you on sight. They hung the killing of that electrician on you."

"That's their way!" Nay Tuman absented gloomily. "A price on my head. They thought I'd stow away on some rocket liner, I suppose."

"Weren't you afraid some desert rat would give you away?"

"No danger. They're just about all fugitives themselves. They hid me till I grew this foliage. They showed me how to find food and water where seemingly there was none. The desert isn't sterile. Why, I know of three or four men within fifty miles of here! Sometimes they stop at my spring for water. As for the harness frames at the fort, those sojers might as well be blind, considering all they miss."

"You asked a while ago if you've changed much. You have. I remember your picture. All of us studied it, because there's a 100,000 I. P. dollar reward out. You were a slim lad then, not the fuzzy bear you are now. How would you like to go in to Tarog with me? They seem to have us licked now—but did you ever hear that the I. F. P. is most dangerous when it's been thoroughly licked?"

"I don't know—I'm used to the solitude," Tuman demurred. "In the city I'd be lost."

But Murray won him over. He had a persuasive way with him.

The next morning they started, guiding their course by the Sun. They made no attempt to travel fast, but the going was easy. Although they rested during the heat of the day, and buried themselves for the nights in the sun-warmed sand, they made about fifteen miles a day. They saw no other human being. These desert dwellers did not meet for mere sociability.

They left the mountains on the second day, descending to the lower level of a broad, sterile plain which was studded by the low, greenish pulp-mounds, that resembled mossy rocks more than vegetation. After two days more they came to a region where huge blocks of stone, of the prevailing orange or brick color, lay scattered around on the plain.

"They look good to me," Tuman said. "If some patrol comes along now we'll have plenty of cover, at least. This belt is a hundred miles wide, maybe a little more. Good hunting there. Plenty of desert hogs, as fat and as round as a ball of bovine butter. I can knock 'em over with a rock, and you can use your neuro, in a pinch."

They did, in fact, succeed in capturing one of the little creatures soon afterward, and, dropping a moistened fire pellet on top of a pulp-mound, soon were roasting their meat.

Not once, however, did either one relax his vigilance. Almost simultaneously they discovered the little black dot that seemed to pop out of the irregular southern horizon. They leaped to their feet, kicked out the fire. They would have covered the ashes with sand but for hundreds of feet in either direction there was nothing but bare rock.

"Never mind!" Murray said. "Let's make for cover. They may think it's an old fireplace. With rains only about once in three years that spot will look like that indefinitely."

"Yes," Tuman agreed, running along, "if they didn't see the smoke!"

As the craft neared they could make out the orange and green of the Martian army.

"From the fort," Murray guessed. "Scar Balta must have had his doubts about me. He ordered them out to finish the job, if necessary."

"It's drifting," Tuman observed. "The driving tail seems to be missing."

"Well, anyway, it's coming down, and where an army ship comes down is no place for us."

They heard the scrape of her keel as she settled down. Murray gave a gasp of surprise.

"Tuman," he muttered, "that fellow wearing the Martian uniform is an I. F. P. agent named Hemingway. The uniform doesn't fit and I bet the man he took it from is no longer alive. Do you know the giant with him?"

"Under that dirt and blood, I'd say he's Tolto, Princess Sira's special pet. No other man of Mars could be that big! Seven or eight years ago—she was just a kid, you know—she picked him up in some rural province. Kids just naturally do run to pets, don't they? And the princess was no exception. But he looks like nobody's pet now. I'd rather have him peg me with his neuro, though, than to take me in his hands!"

They watched as Sime and Tolto slowly walked about in widening circles, and when they were sufficiently far away Murray and Tuman closed in. They had no expectation of finding the ship unlocked, and wasted no time trying to get it. Instead they climbed a flat-topped block of stone about ten feet high. From this position they could command, with Murray's neuro, anyone who might seek to enter the ship.

"These fellows are our best hope," Murray told Tuman. "But we have to convince 'em that we're friends first. Otherwise we're liable to be cold meat, and cold meat can't convince anybody. Keep your head down."

The necessity of lying flat, in order to keep from silhouetting themselves against the sky, deprived them of the opportunity to see. Nevertheless, they could tell, by the sound of their voices, when Sime and Tolto returned. When it seemed that they were directly beneath, Murray risked a look. There they were.

Murray carefully set the little focalizer wheel for maximum diffusion. He felt sure that it would not be fatal, considering the distance and the physical vigor of the men he meant to hold. He pressed the trigger.

"Get down quick!" he snapped. "I'll let up for a second; you grab their neuros."

Tuman executed the order with dispatch. Stepping back, he trained the pistols on their late owners, while Sime and Tolto, a little dazed, stumbled to their feet. A man may argue, or take chances, when menaced by a needle-ray, but mere bravery does not count with the neuros. All men's nervous systems are similar, and when nerves are stricken, courage is of no avail.

CHAPTER IX
PLOT AND COUNTER-PLOT

As these four men faced one another in the slanting rays of the setting Sun far out on the desert, the planetary president, Wilcox, sat in his office in the executive palace in South Tarog, situated, as were so many of the public buildings, on the banks of the canal.

Wilcox was in his sixties. A gray man, pedantic in his speech, his features were strong: his nose, short and straight, somehow, expressed his intense intolerance of opposition. His long, straight lower jaw protruded slightly, symbolizing his tenacity, his lust for power. His eyes, large, gray, intolerant, looked before him coldly. Wilcox was the result of the union of two root-stocks of the human race, of a terrestrial father, a Martian mother. He had inherited the intelligence of both—the conscience of neither.

Now he sat in a straight, severe chair, before a severe, heavy table. Even the room seemed to frown. Wilcox's face was free of wrinkles, yet it frowned too. He seemed not to see the flaming path the setting Sun drew across the broad expanse of the canal, for he was thinking of bigger things. Wilcox was a little mad, but he was a madman of imagination and resource, and he was not the first one to control the destinies of a world.

"Waffins!" His voice rang out sharp and querulous. A servant, resplendent in the palace livery of green and orange, was instantly before him bowing low.

"Who awaits our pleasure?"

"Scar Balta, sire," answered Waffins, bowing low again.

"We will see him."

Waffins disappeared. Scar Balta came in alone, sleek as usual showing no trace of his irritation over his long wait. He did not even glance at the somber hangings that concealed a number of recesses in the wall. Scar knew that guards stood back of those hangings, armed with neuro-pistols or needle-rays as a precaution against the ever-present menace of assassination. And of the loopholes back of these recesses, with still other armed men, as a constant warning to any of the inner guards whose thoughts might turn to treachery.

Scar Balta bowed respectfully.

"Your Excellency desired to see me?"

"I wished to see you, or I should not have had you called," Wilcox replied irritably. "I wish to have an explicit understanding with you as to our proceeding next week at our conference with the financial delegates. Sit here, close to me. It is not necessary for us to shout our business to the world."

Balta took the chair beside Wilcox, and they conversed in low tones.

"First of all," Wilcox wanted to know, "how is your affair with the Princess Sira progressing?"

"Your Excellency knows." Balta began cautiously, "that the news agencies have been sending out pictorial forecasts—"

"Save your equivocation for others!" Wilcox interrupted sharply. "I am aware of the propaganda work. It was by my order that the facilities were extended to you. I am also aware that the princess escaped from Joro's palace. An amazing piece of bungling! Did she really escape or is Joro forwarding some plot of his own?"

"He seems genuinely disturbed. He has spent a fortune having the canal searched by divers, flying ships and surface craft. If Sira fails to marry me Joro's life ambition will fail, for the hopes of the monarchists will then be forever lost."

"True; but his Joro some larger plan? His is a mind I do not understand, and therefore I must always fear. A man with no ambition for himself, but only for an abstract. It is impossible!"

"Not impossible!" Balta insisted. "Joro is a strange man. He believes that the monarchy would improve conditions for the people. And, Your Excellency, wouldn't I be a good king?"

Wilcox looked at him morosely. His low voice carried a chill.

"Do not anticipate events, my friend! There are certain arrangements to be made with the bankers regarding the election of a solar governor!" His large gray eyes burned. "Solar governor! Never in history has there been a governor of the entire solar system. Destiny shapes all things to her end, and then produces a man to fill her needs!"

"And that man sits here beside me, Balta added adroitly. Wilcox did not sense the irony of the quick take-up. He had been about to complete the sentence himself. But his mind was practical.

"The bankers must be satisfied. The terrestrial war must be assured before they will lend their support."

"It is practically assured now," Balta insisted. "Our propaganda bureau has been at work incessantly, and public feeling is being worked up to a satisfactory pitch. Only last night two terrestrial commercial travelers were torn to pieces by a mob on suspicion that they were spies."

"Good!" Wilcox approved. "Let there be no interruption in the work. Our terrestrial agents report excellent results on Earth. They succeeded in poisoning the water supply of the city of Philadelphia. Thousands killed, and the blame placed on Martian spies. Our agents found it necessary to inspire a peace bloc in the pan-terrestrial senate in order to keep them from declaring war forthwith. But these things are of no concern to you. Have you made the necessary arrangements with the key men of the army?"

"I have, Your Excellency. They are chafing for action. The overt act will be committed at the appointed time, and the terrestrial liner will be disintegrated without trace."

"And have you made arrangements for the disposal of the ship's records?"

"Our own ship? I thought it best to have a time bomb concealed aboard. That way not only the records will be destroyed but there will be no men left to talk when the post-war investigating commission comes around."

"Well managed!" Wilcox approved shortly. "See that there is no failure!" He dismissed the young man by withdrawing to his inner self, where he rioted among stupendous thoughts.

Scar Balta emerged into the streets, brightly illuminated with the coming of night, and his thoughts were far from easy. The absence of the princess was a serious handicap—might very easily be disastrous. With her consent and help it would have been so simple! The people, entirely unrealizing that their emotions were being directed into just the channels desired, could most easily be reached through the princess.

First the war, of course, and then, when the threatened business uprising against financial control had been crushed, a planet-wide sentimental spree over the revival of the monarchy and the marriage of the beautiful and popular princess. As prince consort, Scar would then find it a simple matter to maneuver himself into position as authentic king.

But without the princess! Ah, that was something else again! For the first time in his devious and successful career, Scar Balta felt distinctly unhappy. He had schemed, suffered and murdered to put himself in reach of this glittering opportunity, and he would inevitably lose it unless he could find Sira.

In the midst of his unhappy reflections he thought of Mellie.

Sira knew well that Wasil adored her. He had for her the same dog-like devotion that Mellie had. She knew she could ask for his life and he would give it. And what she had planned for him was almost equivalent to asking for his life.

She told him as much, sitting beside him on a bench in the garden. His smooth coral face was alight, his large eyes inspired.

"I will do just as you have commanded me!" he declared solemnly, and would have kissed her hand.

"You must not only do it; you must keep every detail to yourself. You must not even tell Mellie. Do you promise?"

"I promise!"

She kissed him on the forehead. "Farewell, Wasil. I have been here two days already—far longer than prudence allows. They will be here looking for me. Have you any money?"

Wasil produced a roll of I. P. scrip; handed it to her.

"Kiss Mellie for me," she called, as she slipped out of the garden. She was still dressed in the coarse laborer's attire that she had bought on the trading boat, and mingled readily with the crowds in the streets. She hoped she would not meet Mellie, for the girl's devotion might outweigh her judgment.

The rest of that day Sira prowled about the city. Mingling with the common people, she came to have a new insight in their struggles, their sorrows. Passing the walls of her own palace, now locked and sealed, she felt, strangely, resentment that there should be such piled-up wealth while people all around lacked almost the necessities of life.

She surprised herself, also, by a changing attitude toward the life ambition of Prince Joro. The old man's discussions of social conditions that could be corrected by a benevolent monarch had always before seemed to her merely academic and without great interest. Such co-operation as she had given him was motivated entirely by personal ambition. Now she recalled some of Joro's theories, reviewed them in her mind, half consenting.

Always she would strike a barrier when she came to Scar Balta. The more she thought of him the more he repelled her. She puzzled over that. Scar was quite personable.

Tarog, every industrial city along the equatorial belt, and even the remotest provinces, were seething with war talk. The teletabloids at the street corners always had intent audiences. Sira watched one of them. Disease germs had been found in a shipment of fruit juices from the Earth. The teletabloids showed, in detail, diabolical looking terrestrials in laboratory aprons infecting the juices. Then came shocking clinical views of the diseases produced. Men, on turning away, growled deep in their throats and women chattered shrilly. The parks were milling with crowds who came to hear the patriotic speakers.

There was hardly anyone at the stereo-screens, where the news of real importance was given.

"President Wilcox announced to-day that an interplanetary conference of financiers will be held in his office three days from to-day, beginning at the third hour after sunrise. President Wilcox, whose efforts have been unremitting to prevent the war which daily seems more inevitable, declared that the situation may yet be saved unless some overt act occurs." At the same time the device showed a three-dimensional picture of the planetary president, impressive, dominating, stern with a sternness that could mean almost anything.

Sira, hurrying home to an inexpensive lodging house, thought:

"Three days from to-day! I have done what I could. The hopes of the solar system now rest with Wasil. I am only a helpless spectator."

Tarog awaited the conference on the morrow bedecked like a bride. The Martian flag, orange and green, fluttered everywhere. On both sides of the canal the brilliantly lighted thoroughfares were restless with pedestrians, and the air was swarming with taxicabs. Excitement was universal, and business was good.

The glare of the twin cities could be seen far out in the cold desert. Four men, stumbling along wearily, occasionally estimated the distance with wearied eyes and plodded onward.

After a long silence Murray remarked:

"It's just as well that the levitators gave out when they did. We were drifting mighty slow—making practically no time at all. Probably we'd have been spotted if we'd gone much further."

"Yeh?" Sime Hemingway conceded doubtfully. "But they may spot us anyway. We have no passes, and none of us looks very pretty. As for Tolto, we could hide a house as easy as him."

"But we must go on," said Tuman, the Martian. "Yonder lights seem too bright, too numerous for an ordinary day. There's some kind of celebration."

They trudged on for several hours more. Although weariness made their feet leaden and pressed on their eyelids, they dared not halt. Each one nursed some secret dread. Tolto thought of his princess, his child goddess, and mentally fought battle with whomever stood between him and her. Sime and Murray saw in those lights only war, swift and horrible. Tuman imagined a city full of enemies, ruthless and powerful.

Gradually, as they came closer, the lights began to go out one by one. The city was going to bed.

An hour later they came to an illuminated post marking the end of a street. A teletabloid was affixed to this post, buzzing, but its stereo-screen blank. Murray found a coin, inserted it in the slot.

"Marriage of the Princess Sira and Scar Balta will be held immediately after the financial congress," the machine intoned briskly, and in time with its running comments it began to display pictures.

Sime, watching indifferently, caught his breath. It seemed to him that he knew this girl, who appeared to be walking toward him up a stately garden alley. She came steadily forward with a queenly, effortless stride. And now it seemed as if she had seen him, for she turned and looked straight into his eyes. It seemed that her expression changed from laughing to pleading. And he recognized the girl with the stiletto whom he had caught in his hotel room.

He said nothing, however. He could hardly explain the feeling of sadness that came over him. He stood silent, while the others commented excitedly over the overshadowing war news.

"It's all in the box," Tuman said gloomily. "Many times I've helped cook up something like this. The boys in the central offices are laughing, or swearing, as the cast may be. The poor devils don't own their own souls, if they're equipped with any. I'd rather be here, expecting to be thrown into a cell by daylight!" He shivered in the night chill.

They ran into a little luck when they needed it most. A roving taxi swooped down upon them, hailed them for fares. They flew the rest of the way in. Their luck held. A city policeman, noting their stumbling walk as they lurched into a cheap hotel, did not trouble them for their passes. He had seen many such men that night, soldier and civilian, with clothes bloody and torn. The excitement of the day, coupled with the fact that nearly everyone carried arms, had led to numerous fights, not a few of which ended fatally.

"Merclite!" grinned the policeman, suppressing a hiccup of his own. "And besides, that big 'un would make two of me."

CHAPTER X
ONE THOUSAND TO ONE

The scheme that Sira had imparted to Wasil was simple—simple and direct. Moreover, it was sure, provided it succeeded. Its execution was something else again. Its chances were, mathematically expressed, about as follows:

If every single detail worked as expected, a great and smashing success. Ratio: 1:1,000.

If one single detail failed, immediate and certain death for Wasil. Ratio: 1,000:1.

The princess knew that the power of Wilcox, his supporting oligarchy and the interplanetary bankers, was all based on the skilful use of propaganda. If the people of Mars and of Earth knew the forces that were influencing them, their revulsion would be swift and terrible. There would be no war. There would be events painful and disastrous to their present rulers, but a great betterment of humanity's condition.

The key to the situation was the news monopoly, the complete control of all broadcasting—of the stereo-screens, the teletabloids—that colored all events to suit the ends of the ruling group. The people of Mars as well as of Earth were capable of intelligent decision, of straight thinking, but they rarely had an opportunity to learn the truth.

They had now, by a knowing play on their emotions, directed by psychologists, been wrought to a point of frenzy where they demanded war. Their motives were of the highest in many individuals—pure patriotism, the desire to make the solar system safe for civilization. The bright, flaming spirit of self-sacrifice burned clear above the haze and smoke of passion.

What would happen if all these eager millions of two neighboring planets were to learn the true state of affairs? Sira knew what transpired in those secret conventions, when double guards stood at all doors and at the infrequent windows; when all communication was cut off and the twin lenses of the telestereos and the microphones were dead. Prince Joro had told her, with weary cynicism. But Joro had also told her that the oligarchs guarded this vital and vulnerable point with painstaking care.

Sira had reached inside their first defense, however. Wasil was loyal to his salt, but he had both loyalty and affection for Princess Sira. As the day of the interplanetary financial conference leaped into being, he was on his way to the executive hall that lay resplendently on the south canal bank, ready to lay down his life.

The hall proper was really only the west wing of the magnificent, high-arched building. Its brilliant, polished metal facade reflected the light of the rising Sun redly. The east wing, besides housing various minor executive offices, also contained the complicated apparatus for handling the propaganda broadcastings. On the roof, towering high into the air, was a huge, globular structure, divided into numerous zones, from which were sent various wave bands to the news screens both on Mars and on Earth. The planetary rulers had taken no chances of tampering with their propaganda. The central offices, where news and propaganda were dramatized, were in another building, but as everything from that source had to pass the reviewing officer, a trusted member of the oligarchy himself, in his locked and guarded office, this did not introduce any danger of the wrong information going out to the public.

When Wasil reached the broadcasting plant, he was admitted by four armed guards. He locked the door behind him, to find his associates already busy, testing circuits and apparatus. Stimson, the chief engineer, was sitting at his desk studying orders.

A few minutes later he called the men to him. There were three others besides Wasil: young Martians, keen, efficient, and, like most technies, loyal to the government that employed them.

"Sure are careful to-day," Stimson grunted, scratching his snow-white hair, which was stiffly upstanding and showed a coral tinge from his scalp. "Must be mighty important to get this out right. Wilcox personally wrote the order. If any man fumbles to-day, it's the polar penal colony for him!" The Sun-loving old Martian shivered.

"And here's another bright idea. Only one man's to be allowed in the plant after the circuits are all tested! How'n the name of Pluto will he handle things if a fuse blows? But what do they care about that! We're technies! We're supposed to know everything, and never have anything go wrong!"

"But why only one man?" cried Scarba, one of the associate engineers. "It's asking too much! I'll not take it on, far as I'm concerned. My resignation will be ready soon's I can get a blank!"

"I too! I'm with you, Scarba!" "We work like dogs to get everything in first-line condition, and then—" The hard-working and uncomplaining technies were outspoken in their resentment.

"Oh, I see your point," Stimson agreed. "I could stand Balta, but Wilcox is just one too many for me. But do you boys think for one minute we could get away with a strike?" He laughed angrily. "I can remember when the technies were able to demand their guild rights. But you boys weren't even born then. Now, let's get this straight:

"We are going to do just as we are told. Wilcox, of course, never explains an order, but the reason for having only one operator on the job is simply to concentrate responsibility on that one man. There will be no excuse if he fails. Before the convention starts, and after it is over, there will be a message to send out. The convention itself will be secret, as usual. During the convention, there will be some kind of filler stuff from the central office."

"Yeh!" snorted one of the men. "That's the dope, all right. One of us is stuck, but if it's me I'll walk out and head for the desert."

Stimson looked at him with a sardonic smile. "I forgot to mention: the doors will be locked and barred, and of course there's no such thing as windows."

Wasil whistled. "They're sure careful. Well, Stimson. I haven't a thing to do all day. I'll take it on."

They all looked at him, not sure that they had heard him right.

"What's the matter, sonny?" Stimson said slowly. "Too much Merclite last night? You're shaking!"

"It's an opening!" Wasil insisted.

"An opening to tramp ice at the pole for the rest of your life!"

"All right. I'll chance it!"

They consented, without very much argument, to let Wasil have the dangerous responsibility. At 2:30, two and a half hours after sunrise by the Martian reckoning, he signed a release acknowledging all circuits to be in proper order, and was locked behind the heavy doors, alone with a maze of complicated apparatus and cables that filled the large room from floor to ceiling.

Now it was done! Chance had thrown Wasil into a position where he could, without great danger of failure, carry out his plan. But at the same time things had so fallen that he, Wasil, must now die, regardless of the outcome!

If he succeeded in broadcasting the proceedings of the convention, and if they had the effect of arousing the public against Wilcox, there would still be no escape for Wasil. Wilcox, or Scar Balta, would come straight for this prison, neuro-pistol or needle-ray in hand!

Even if he should fail, death would be his portion for the attempt.

So thinking, Wasil sat down and carefully re-checked the circuits. The filler broadcast from central office must be sent to the twin cities of Tarog. Otherwise the convention would learn too soon what was happening, and would interrupt its business. The thousands who waited outside on the broad terraces must be regaled with entertainment, as had been originally planned.

But as for the rest of Mars, and Earth, they would get the truth for once. Those bankers would speak frankly, in the snug isolation of the hall. No supervision here. Conventions, empty politeness, would be forgotten. Sharp tirades, biting facts, threats, veiled and open, would pass across the table between these masters of money and men.

But this time they would be pitilessly bared to the worlds!

Feverishly, Wasil inspected the repeater. It was a little-used device that would, an hour or two later, as desired, give out the words and pictures fed into it. Although Tarog would not learn the convention's secrets as quickly as the rest of Mars, or Earth, Tarog would learn. Wasil threw over the links and clamped down the bolts with a grunt of satisfaction. When a man is about to die, he wants to do his last job well.

Suddenly a red light glowed, and a voice spoke.

"Special broadcast. Tarog circuit only!"

"Mornin', Lennings," Wasil remarked to the face in the screen. "All set? Go ahead."

The central office man held up a thick bundle of I. P. scrip, smiled pleasantly, saying:

"Somebody in North or South Tarog, or in the surrounding territory, is going to be 100,000 I. P. dollars richer by to-morrow. How would you like to have 100,000 dollars? You all would like this reward. It represents the price of a snug little space cruiser for your family; a new home on the canal; maybe an island of your own. It would take you on a trip to the baths of Venus and leave you some money over. Of course you all want this reward!

"Now, if you'll excuse me a moment—"

The man's picture faded, and the screen glowed with the life and beauty of Princess Sira—Sira, smiling and alluring.

"You all know this young lady," the announcer's voice went on. "The beloved and lovable Sweetheart of Mars, the bride of Scar Balta—"

The Martian's sleek and well-groomed head appeared beside that of the girl.

"—Scar Balta, whose services to Mars have been great beyond his years; who, in the threatening war with Earth, would be one of our greatest bulwarks of security."

The announcer's face appeared again, stern and sorrowful.

"A great disaster has befallen these lovers—and all the world loves a lover, you know. Some thugs, believed by the police to be terrestrial spies, have kidnapped the princess from the palace of her uncle, Prince Joro of Hanlon. It is believed that they had drugged her and hypnotized her, so that she has forgotten her duty to her lover and her country."

The green light flashed, and Wasil broke the circuit. The central man lingered a moment, favoring Wasil with a long wink.

"What a liar you're getting to be!" Wasil remarked coldly. But the central man, not offended, laughed.

So they were offering a reward! And urging further treachery as an act of patriotism! Wasil was not too much excited, however. The disguise the princess had chosen would probably serve her well. Besides, she had promised to keep in retirement as much as possible.

Clack! Clack! The electrically controlled lock of the door was opening. Only Wilcox knew the wave combination. Wasil felt a chill of apprehension as the door opened and Scar Balta strode in. He was fully armed, dressed in the military uniform; but the former colonel was now wearing on his shoulder straps the concentric rings denoting a general's rank.

CHAPTER XI
GIANT AGAINST GIANT

Although Princess Sira had promised to keep out of the way, she could not resist the powerful attraction of the executive hall, in which, on this day, the fate of two planets was to be decided. As the crowds of people began to drift toward the hall, she joined them, still dressed in her laboring man's shapeless garments, the broad sun-helmet hiding her face effectively. Her long, black hair was concealed under the clothing. Having nearly been drawn into a brawl the day before, she now carried a stained but still very serviceable short sword that she had purloined from a merclite-drunken reveler in a gutter.

Thousands were already on the terraces surrounding the government buildings. They were milling about, for it was still too soon after the night's chill to sit down or lie on the rubbery red sward. Taxis were bringing swarms over the canal from North Tarog, and water vehicles were crossing over in almost unbroken lines.

Already the merclite vendors were busy, making their surreptitious way from group to group, selling the highly intoxicating and legally proscribed gum that would lift the users from the sordid, miserable plane of their daily existence to exalted, reckless heights.

War vessels now began to course overhead, their solid, heavily plated hulls glinting dully in the sun. Their levitator helices moaned dismally, and as their long, slanting shadows slid over the assembled thousands, it seemed that they cast a prophetic pall; that there was a hush of foreboding.

But the psychological expert high in a nearby tower immediately noted the slump in the psycho-radiation meter whose trumpet-shaped antenna pointed downward. At the turn of the dial the air was filled with throbbing martial music, and the expert noted with contemptuous satisfaction that the needle now stood even higher than before.

Sira, caught like all the rest of the people in that stirring flood of music, felt her own pulse leap. But she thought:

"This is the day! Wasil, could I only be with you!"

She thought sadly of Joro, whose shrewd observations and counsel she missed more than she had ever thought possible.

"Poor, dear Joro! You would be a better king than any man you could ever find! I wish I could have done as you wished me to."

There was a stir near the main entrance of the hall. A large private yacht was slowly descending. She was bedecked with the green and gold bunting of the terrestrial government, the green and orange of Mars. Her hull glittered goldenly.

"Back!" shouted the captain of a Martian guard detail, the soldiers running with pennant-decked ropes looping after them. The crowd surged against the barrier, but more guards were sent out as reinforcements, until they had cleared a space for the ship and a lane to the hall entrance.

"Mars greets the distinguished guests from our sister planet!" boomed the giant loudspeaker in the tower. Immediately afterward came the strains of the song—"Terrestria—Fair Green Terrestria"—in a rushing torrent of sound. But the frank and fluent melody was strangely distorted, with unpleasant minor turns and harsh whisperings of menace, and the tower psychologist noted a further rise of the needle.

There was a diversion of interest now. The mob of first arrivals, as well as the ever-freshening stream of newcomers, was moving toward the teletabloids and the more conservative stereo-screens. On this occasion they were both carrying the same message, however. Sira heard the propaganda division's latest fabrication about her alleged kidnaping by terrestrial agents. She needed no radiation meter to tell her of the intense wave of hatred for the Earth that swept over the densely packed area. And this was followed by another emotion—a wave of cupidity—set up by the offer of 100,000 I. P. dollars reward for her return. She saw about her faces greedy, faces wistful, even compassionate faces. But outnumbering them by far were faces set in truculent mold.

Sira moved restlessly from place to place, feeling more deeply depressed with every moment. She felt as if she had been left entirely out of life, friendless, alone. Among all these thousands she had no friend. It seemed to her that never before had there been such a paucity of monarchists. Sharp-featured, with a wire-drawn manner of efficiency and resolution about them, they had constituted almost another race among this practically enslaved people, maintaining for themselves a tolerable position despite the opposition of the oligarchy. Now, however, they seemed to have vanished. All that morning Sira had not seen one. She would not have disclosed her identity, but it would have been comforting to see one of those friends of old.

She was stopped by a jam. Looking between the bodies of two large and sweaty men, she realized that someone was standing on a surveyor's marking block, delivering a speech.

"The great Pantheus has so decreed it," the speaker was shouting in a cracked voice that at times dribbled into a whine. "We must shake off forever this menace from the green planet—this planet dominated by wicked women.

"Oh, my friends, last night they came to me in dreams, these pale women of the green star. They tempted me and they mocked me. They laid their cold hands on my throbbing brow, and their cold hands burned me!

"Oh great Pantheus! How I have suffered! The creatress who in her malice created this wicked world beyond the gulf—"

The Martians were entertained by the quavering denunciation. Some grinned broadly at one another; others placed their thumbs in their ears and wiggled their fingers. But the old man continued. Finally, two of the foremost spectators, sensing the tiny body crowded between them, stepped aside.

"Don't miss this, my little man. Listen, and maybe you will laugh yourself a little bigger." He gave Sira a gentle shove, so that she almost stumbled over the block on which the speaker was standing.

And that old man suddenly stopped talking, so that his toothless mouth sucked in, then stood agape. The rheumy eyes rolled, and a wisp of dirty gray hair strayed across his gnarled face. He lifted a shaking hand, pointed a knotty finger.

"There she is!" he croaked. "There she is! I claim—"

"There she is!" guffawed a tipsy merclite chewer. "The creatress, come to punish you! Cut off his nose, O creatress, and stuff it into his mouth!"

There were shouts of laughter, a surge to see better.

"No! No! I, Deacon Homms, claim the reward!" the old man screamed. "She is the princess; I know her. She came out of the canal to tempt me! She is the Princess Sira. Now shall I at last enter the Palace of Joys! I claim the 100,000 dollars!"

But he still had to catch Sira. The crowd, suddenly sensing that this old fanatic might be telling the truth, rushed in savagely, each eager to seize the prize, or at least to establish some claim to a share of the award. Men and women went down, to be trampled mercilessly. Inevitably they got in one another's way, and soon swords were rising redly, falling again.

"Guards! Guards! A riot!" Some were fleeing the scene; others rushing in, grateful for the opportunity to expend excess pugnacity. A fresh platoon of soldiers tumbled out of a kiosk leading to an underground barracks like ants out of a disturbed nest. They deployed, holding their neuro-pistols before them, focalizers set for maximum dispersion, therefore non-fatal — merely of paralyzing intensity. Some of the rioters now turned to run, but others persisted, willing to be rendered unconscious, just so it would be near the valuable princess.

A few moments later the captain of the guard surveyed the mass of paralysed bodies and the sword-slashed corpses, all intermingled.

"What's this all about?" he demanded of a scarred, evil-looking fellow who was the first to rise to his elbow.

"The Princess Sira! I claim the reward. In there! She stood right there!"

"Get out, you galoon!" the captain growled, knocking the fellow unconscious with the heavy barrel of his neuro. "Sort 'em out there. Moggins, Schkamitch. On the double. You will share, according to rank."

But eagerly as they searched, they did not find Sira. Creeping between the legs of the maddened reward seekers, she had fought clear, had gained the shelter of a tall, red conical tree whose closely laced branches pressed her to the ground, clinging to the greasy trunk.

She realized that her sanctuary was none too secure. There would surely be a methodical search after the first excitement, and she would be discovered. She had lost her sun-helmet, but nevertheless she must risk making a break. A large proportion of the people were wearing such helmets. Perhaps she could snatch one.

But before such an opportunity came, she saw a chance to dash to a nearby clump of shrubbery. On the other side was a long hedge, leading to an alley back of a group of warehouses. If she could gain this alley, she felt sure she would be safe for the time being.

All over the park, which was thirty or forty acres in extent, there were minor riots, as some unfortunate was mistaken for the princess and blindly struggled for.

Sira lost no time. She scuttered along the hedge like a frightened kangrat. But as she crossed a small open space, a stentorian voice shouted:

"There she is! That's her! The princess!"

Out of the corner of her eye she saw him, running toward her lumberingly, his great arms outspread. Tuman had been wrong in saying that on all of Mars there was no man as big as Tolto. This one was, and he looked more formidable. Instead of Tolto's normally good-natured face, this one looked like an enraged terrestrial gorilla, although at the moment it was really expressing joy and eagerness.

Several other men joined the chase, and then scores. They were fleeter of foot than the ape-man, but as they passed him in the narrow alley he smashed them to the pavement with casual blows of his terrifying hands. Thereafter he was undisputedly in the lead; the others content to follow in his rear, although many were armed, and the giant was not.

This was an advantage to Sira. The whole mob was slowed by the lumbering pace of the ape-man, and she was able to keep in the lead without difficulty. Several times some of her pursuers ran ahead by other routes, intent on snatching her into some doorway. But each time she slashed at them with her sword, springing past.

She had not run very far when her fear of another danger was realized. There was a high, keen whistle overhead, and a scouting police car flashed near. Under the neuro-pistols both hounds and hare would be paralyzed, and she would be easily taken. Sira longed for one of these handy weapons herself, but they were too expensive: she had been unable to secure one.

Now the police car was coming back. The sliding forward door was drawn back, and a man was leaning out, neuro alert. Judging the distance expertly, he pulled the trigger, and a hundred men fell unconscious.

"Got 'em!" he snapped over his shoulder. "The princess as well. Down quick!"

Sira, spared because of the officer's unwillingness to take a chance on injuring her, leaped through a gap in a wall and sprinted through a garden smothered with thick, leathery-leaved weeds, some of them higher than her head. She almost laughed with relief, but as she flitted around the corner of a house toward the street she saw the gorilla faced giant again in pursuit, and beyond the garden wall the police ship was just settling to the ground.

It just seemed to be raining giants that day. Sira ran out of a narrow gate at the front of the house into the street, to be stopped by a tremendous human framework as solid and unyielding as a mountain. She stepped back, drew her sword—

"Softly! Softly!" a rumbling bass implored. "Doesn't the Princess Sira recognize her servant, Tolto?"

"Tolto!" All at once the tautness went out of her, and Sira leaned against the wall, divided between laughing and crying.

"Tolto and his good friends were looking for you," the big man rumbled anxiously. "The teletabloids said there was a riot coming—"

He got no further. The gorilla-faced pursuer catapulted himself sideways through the portal, being too wide to go through in the regular way. He emitted a raucous shout of triumph:

"I got her! It's her, all right! I claim—"

As he reached out his enormous sun-blackened arm there was a thud that seemed to shake the ground. Instantly enraged, the man's little red-rimmed eyes jerked quietly to the dealer of that shocking blow. Then the conical little head sank between the bulging shoulders, the long, thick arms bowed outward, and the ape-man launched himself at Tolto.

That was a battle! On the one side devotion, simple-minded loyalty and a fighting heart in a body of such mechanical perfection as Mars had never seen before or since. On the other side a primal beast, just as huge, rage-driven, atavistic, savage.

Fists as large as an average man's head, or larger, crashed against unprotected face and body. Gigantic muscles rippled and crackled. Blows echoed from wall to house and seemed to thud against the hearts of the spectators.

It was as if time and memory had come to a standstill. The present was not, nor present ambitions and duties. The soldiers came plunging out into the street, swords in their hands, but they stopped to watch. Sime, Murray and Tuman, used to instant and automatic battle, watched. A struggle so titanic, by tacit, by unconsidered consent, must be left to decide its own course.

Tolto seemed to be slowly gaining an advantage. During his novitiate as a palace guard the other men had instructed him in the science of their pastime-fighting. Although he scorned to guard against the blows of his savage antagonist, he placed his own punches more shrewdly, more effectively. The ape-faced one, through a red film, sensed that he was being beaten, and that this fight would end in death.

Suddenly he changed his tactics. Rushing in, he threw his arms around Tolto's great torso. He opened his jaws, and his long yellow fangs bit into the flesh of Tolto's shoulder.

Tolto, taken slightly by surprise, met this new menace promptly. Placing his powerful forearm against the battered, hairy face, he attempted to bend the head back. But it was so small, in proportion, and so slippery with blood, that he was unable to dislodge it.

So Tolto matched brute strength against brute strength. His arms encircled his enemy's body, and the tremendous muscles of his shoulders and body began to arch.

So they stood poised for a few seconds, as if on the brink of eternity.

"Go-o-o-wie!" exclaimed one of the soldiers, awed.

Slowly, like the agonizingly slow plastic creep of metal under great pressure, the gorilla-faced giant was yielding. His dark skin became mottled. His breath came gaspingly. His rope-knotted arms slipped a little.

But it was not in him to surrender, which might still have saved his life. With a vicious twisting motion of his head he tried to drag his fangs through the thick muscles of Tolto's shoulder. The wound began to bleed more freely, choking the savage at each labored breath.

Now Tolto began to walk forward. Always his antagonist had to yield a little, unwillingly, grudgingly, just enough to keep the paralyzing pressure on his spine from becoming unbearable. And slowly, inexorably, Tolto followed. His arms tightened. His leg slipped suddenly between the ape-faced man's supports. Tolto grunted. The sound seemed to labor upward from his innermost being, his body's protest as he called upon it for its last reserve of strength.

Like an echo, there was a dull crack, a brief, agonized moan from the ape-faced one; and the savage, unknown giant slumped to the pavement, dead with a broken back. Tolto staggered to the wall, breathing deeply.

"Man, what a fight! What a *fight!*" The young Martian captain passed a shaking hand over his face. The battle had stirred him more deeply than he wanted to admit. But in a few seconds he came out of his mental maze.

"Attention! All right, men, you're under arrest. As for the girl—"

"As for the girl," came a clear feminine voice, as Sira stepped out from the shelter of a buttress some dozen feet away, "—the girl took advantage of your preoccupation to relieve you of your neuros. As you see I have two of them in my hand. The rest of them are over by that wall. No! Don't try to rush! You are welcome to your swords, but they are useless here."

CHAPTER XII
"HE MUST BE A MAN OF EARTH"

Friend and foe looked stupefied. But they were used to the give and take of battle. That this girl should disarm a detachment of soldiers while they and their own men were absorbed in such a common thing as a fight struck them as humorous. They laughed.

"This is a better break then we deserve," Sime said, grinning with a trace of sheepishness. "Captain, you take your men across the street and hold 'em there. We're going to borrow your car. No funny stuff!" Civilians were flooding into the streets. There would soon be a mob.

"We will not," replied the captain, "try any funny stuff. Some day, my friend, I hope to open you up with my sword," he added.

"By all means," Sime agreed pleasantly. "My time is pretty well occupied, but there's no telling when I may meet you again, in my business. Good day, Captain!"

Tuman stayed at the front gate with his neuro while the others struggled through the weedy garden to the police ship in the alley, rejoining them as they were ready to rise.

A crowd had gathered. If they wondered at the appearance of these ragged, scarred and bewhiskered men; at sweat and blood-covered giant Tolto; the obviously high-bred girl in the laboring man's garments, they wisely refrained from comment or action, in deference to the neuros with which the party was bristling.

Once inside and safely in the air, they had time to breathe. Murray, with a gallantry that sat ill on the scarecrow figure he was, cleared matters up a trifle.

"Princess Sira? As I thought. Princess, or Your Highness, to be formal, I am your humble and disreputable servant, Lige Murray, of the Interplanetary Flying Police. Likewise this gentleman behind the brush— Sime Hemingway. You know Tuman? You've missed something, Your Highness! And Tolto! Lucky man!"

Sira recovered quickly from her reaction following the fight. She found a first-aid kit, bandaged Tolto's wounded shoulder skilfully and quickly. She had given no sign of recognition as Sime awkwardly bowed, during Murray's introduction, but now, as Sime held a roll of bandage for her, she gave him a sidewise look, agleam with mischief.

"But I have decided to remit the punishment—the sentence I passed on you, Mr. Hemingway," she said, her sweet, child-like face innocent.

"What punishment?" Sime gasped.

"Why, the punishment of death! For kissing me that night!" she laughed, turning her back.

Murray was heading back for the government park. It was a short distance with the police car. Soon the broad grounds, with their scattered, magnificent buildings, lay below them. But the parks were strangely bare of living creatures. Here and there lay the bodies of men or women.

"Something's happened!" Murray shouted excitedly. "Look out!"

He swerved the ship sharply. They escaped damage as an atomic bomb, unskilfully aimed, exploded far to one side.

"Funny thing, firing on a police car," Sime puzzled. "They might have got news from that detachment we grounded, but how do they know this isn't some other police or military car?"

"Those aren't soldiers," Murray decided. "There's been a riot, and some civilian's got hold of an ato-projector."

"I know what's happened!" Sira exclaimed suddenly. "Wasil—a technie—has managed to broadcast the secret session! That upset their psychology. Oh!" Her face was alight, and she threw up her arms in ecstasy. As quickly she subsided, and tears came to her eyes.

"Wasil!" she cried. "If he is dead, Mellie will never forgive me!"

"Where is this technie?" Sime asked bruskly.

"In the broadcast room. But they have probably killed him."

"Never can be sure. Head her smack for the main entrance, Murray!"

Murray threw the car into a steep dive, and the hall portal rushed up to meet them. A soldier came partially out of concealment, waved a signal. Murray paid him no heed.

They struck with a crash. The stout car crushed through the glittering doors of metal and glass, and before the fragments fell the four men were in the thick of short, sharp and decisive battle. Their neuros hissed venomously, spanged as they met opposing beams. And the princess, struggling through the wreckage, wept tears of rage as the coarse fabric of her clothing caught, entangled hopelessly, and held her.

"Something queer!" Murray said, as they halted for breath after routing what little opposition they had encountered. "Maybe it's a trap. But what an expensive trap for somebody! Where's this broadcasting plant?"

"This way!" Tuman called eagerly. "Maybe we can still save the poor fellow who turned the trick. Broadcast the secret sessions! Don't tell me that little girl isn't fit to rule!"

The heavy metal doors were open, and they hurried in. But Tolto, noting that the princess had not followed, hurried out in search for her.

Sime stumbled over a body. It had been a dark, sleek, youngish man. A jagged burn on his throat told of the needle-ray. "Who's this fellow, Murray?"

Murray glanced at the body. He smiled a brief smile of satisfaction.

"That's Scar Balta. Got what's coming to him at last. Help me with this bird: he's still alive. Cold, though!"

"Got a shot of neuro. Could this be the technie?"

Sime found a fountain of water. He filled a cup, dashed it over the still face. The shock made the man's lips move.

"Mellie, I did it!" he whispered.

"Who's Mellie?" Sime asked.

"Mellie? Seems to me the princess mentioned her name, This is her brother. He's the right guy! Take it easy, brother!"

But Wasil was able to sit up.

"I sure fooled him!" he gasped. "Mixed up the circuits. Scar Balta sat right here while I broadcast the secret sessions, and he was watching a lot o' haywah in the control screen.

"When Wilcox got word from outside he knew he was done. He thought Scar'd double-exed him, so came here in person and gave him the needle-ray."

Despite his nausea, Wasil looked happy.

"Wilcox tried for me, but I dodged back of those frames. So he tried for me with the neuro. The mob was getting wild outside; there was—"

He could not finish. There was an explosion that shook the building to its foundations. Tolto came running in. Sira close after him:

"Joro is coming. Joro has detonated the warships. The hall guards have surrendered. The council is locked up. It can't escape!"

Events were transpiring too fast for comprehension. It was several days later, on a bench in Prince Joro's palace grounds, that Sira summed it up for Sime Hemingway.

"I'm going to accept the throne!" she said. "I'm going to be a real queen. Joro has convinced me that it will be a real service to Mars. The dear old man has schemed and worked so long, so unselfishly."

"Yeh, and he wasn't afraid to fight!" Sime added admiringly. "When he came charging out of those ships with his gang of monarchists, swords flashing, it was a pretty sight to see. And when they closed in on that gang of cheap politicians! Talk about rats in a corner!"

"The prince can fight with his brains as well as with his sword." Sira submitted. "The whole thing would have been hopeless, if he hadn't invented the detonating ray that disposed of the warships. You remember those heavy explosions, shortly after we dropped in the hall, as one might say? Those were the last of them."

A silence fell between them, and Sime was now conscious of the fragile-seeming, so deceiving beauty of this Martian girl. Something had come between them, stripped away the masculine frankness that had existed during their short and dangerous time together. Perhaps it was the softly revealing drape of the thread-of-gold robe she was wearing—true queenly garb, donned by her for the first time.

"There is one requirement that Joro insists on," Sira said in a low voice.

"What's that?" asked Sime, marveling that such transparently pink fingers should handle a sword so well.

"He says that I must choose a mate, to insure the stability of the royal house."

It seemed to Sime that this announcement gave him a pang out of all proportion.

"That should be easy," he managed. "Every Martian is crazy about you."

"He may not be a Martian. He must be a man of Earth," Sira stated firmly.

"Is that so?" Sime asked, genuinely surprised. "Why does Joro insist on that?"

"It is not Joro who insists. It is myself."

Sime found himself looking into eyes filled with shy pleading. He could not, would not, for all of the solar system, have committed the unpardonable affront of rejecting the love so frankly offered. And yet he did not know how to accept this miracle. He did it clumsily, haltingly disclosing the secret recesses of his own heart and what had transpired there since the night he had taken the knife away from her and kissed her.

www.ingramcontent.com/pod-product-compliance
Lightning Source LLC
LaVergne TN
LVHW041738190726
843493LV00008B/2412